Lecture Notes in Mathematics

Editors:
J.-M. Morel, Cachan
F. Takens, Groningen
B. Teissier, Paris

FONDAZIONE CIME
ROBERTO CONTI

CENTRO INTERNAZIONALE MATEMATICO ESTIVO
INTERNATIONAL MATHEMATICAL SUMMER CENTER

C.I.M.E. means Centro Internazionale Matematico Estivo, that is, International Mathematical Summer Center. Conceived in the early fifties, it was born in 1954 and made welcome by the world mathematical community where it remains in good health and spirit. Many mathematicians from all over the world have been involved in a way or another in C.I.M.E.'s activities during the past years.

So they already know what the C.I.M.E. is all about. For the benefit of future potential users and co-operators the main purposes and the functioning of the Centre may be summarized as follows: every year, during the summer, Sessions (three or four as a rule) on different themes from pure and applied mathematics are offered by application to mathematicians from all countries. Each session is generally based on three or four main courses $(24-30$ hours over a period of $6-8$ working days) held from specialists of international renown, plus a certain number of seminars.

A C.I.M.E. Session, therefore, is neither a Symposium, nor just a School, but maybe a blend of both. The aim is that of bringing to the attention of younger researchers the origins, later developments, and perspectives of some branch of live mathematics.

The topics of the courses are generally of international resonance and the participation of the courses cover the expertise of different countries and continents. Such combination, gave an excellent opportunity to young participants to be acquainted with the most advance research in the topics of the courses and the possibility of an interchange with the world famous specialists. The full immersion atmosphere of the courses and the daily exchange among participants are a first building brick in the edifice of international collaboration in mathematical research.

C.I.M.E. Director
Pietro ZECCA
Dipartimento di Energetica "S. Stecco"
Università di Firenze
Via S. Marta, 3
50139 Florence
Italy
e-mail: zecca@unifi.it

C.I.M.E. Secretary
Elvira MASCOLO
Dipartimento di Matematica
Università di Firenze
viale G.B. Morgagni 67/A
50134 Florence
Italy
e-mail: mascolo@math.unifi.it

For more information see CIME's homepage: http://www.cime.unifi.it

CIME's activity is supported by:

- Ministero dell'Universita', Ricerca Scientifica e Tecnologica COFIN '99
- Ministero degli Affari Esteri - Direzione Generale per la Promozione e la Cooperazione - Ufficio V
- CNR Consiglio Nazionale delle Ricerche
- E.U. under the Training and Mobility of Researchers Programme
- UNESCO-ROSTE, Venice Office

Mitchell A. Berger • Louis H. Kauffman
Boris Khesin • H. Keith Moffatt
Renzo L. Ricca • De Witt Sumners

Lectures
on Topological
Fluid Mechanics

Lectures given at the
C.I.M.E. Summer School
held in Cetraro, Italy
July 2–10, 2001

Editor:
Renzo L. Ricca

 Springer

FONDAZIONE
CIME
ROBERTO CONTI

Editor

Renzo L. Ricca
Department of Mathematics
and Applications
University of Milano-Bicocca
Via Cozzi 53
20125 Milano, Italy
renzo.ricca@unimib.it

Authors: see List of Contributors

ISSN 0075-8434   e-ISSN: 1617-9692
ISBN: 978-3-642-00836-8   e-ISBN: 978-3-642-00837-5
DOI: 10.1007/978-3-642-00837-5
Springer Dordrecht Heidelberg London New York

Library of Congress Control Number: 2009926297

Mathematics Subject Classification (2000): 37Bxx, 57Mxx, 57M25, 76Bxx, 76W05

Cover illustration: The illustration shown on the cover (adapted by R.L. Ricca) is taken from the article "DNA knots reveal a chiral organization of DNA in phage capsids", by Arsuaga et al., PNAS vol. 102 (26), 9165–9169. Copyright (2005) National Academy of Sciences, U.S.A.

Cover design: SPi Publisher Services

Printed on acid-free paper

Springer is part of Springer Science+Business Media (www.springer.com)

Preface

It seems very appropriate to publish these *Lectures* in the 150th anniversary of the publication of Helmholtz's seminal paper on vortex motion (1858), that may be regarded as the pioneering work on fundamental questions in topological fluid mechanics. The field is going through a period of great revival, benefitting from the formidable progress in knot theory and differential topology, on the one hand, and mathematical and computational fluid dynamics, on the other. It is therefore with great pleasure that I warmly thank the contributors to this volume for providing such interesting collection of valuable research papers. All the material presented here is actually an update on the original material presented in the lecture notes delivered by six of us, on the occasion of a CIME Summer School (of the Unione Matematica Italiana) in 2001, a School that I had the honour and pleasure to organize and direct in Cetraro, a charming location on the rugged coastline of southern Italy.

The Summer School was qualitatively very successful, thanks also to the many attendees, who themselves were renown experts in their own fields, coming from UK, Germany, Russia, USA, Japan, Tunisia, Ukraine, Canada and, of course, Italy! Given this luxury, as Editor-in-charge of the proceedings, I couldn't resist to break one of CIME's most consolidated rules, by inviting one of the attendees, *Patrick Bangert*, as a representative of such a qualified audience, to collect some background information for an introductory chapter to knot and braid theory. The material of **Braids and knots** (71-page long!) is therefore an invaluable, complementary addition to the lecture programme of 2001, covering elements of braid theory and knot polynomials, as well as more advanced aspects of braid and knot classification, including up-dates on the Word Problem, the Conjugacy Problem and Markov's Theorem. **Topological quantities: calculating winding, writhing, linking and higher order invariants**, by *Mitch Berger*, is on winding number techniques to calculate topological numbers for both closed and open curves, with applications to magnetic fields in magnetohydrodynamics. **Tangles, rational knots and DNA**, by *Lou Kauffman* and *Sofia Lambropoulou*, is dedicated to the theory of tangles and unoriented and oriented rational knots, with applications to DNA recombination processes. **The group and Hamiltonian descriptions of hydrodynamical systems** is due to *Boris Khesin*, who reviews applications of Hamiltonian approach and group theory to ideal fluid dynamics and integrable systems, with particular emphasis on the Landau-Lifschitz and Korteweg-de Vries equations. *Keith Moffatt* presents and discusses in **Singularities in fluid dynamics and their resolution** three types of singularity that can arise in fluid dynamical problems: (i) singularities driven by boundary motion in conjunction with viscosity; (ii) free-surface (cusp) singularities associated with surface-tension and viscosity; (iii) interior point singularities of vorticity associated with intense vortex stretching. In paying tribute to 150 years of topological fluid mechanics, *Renzo Ricca* (**Structural complexity and dynamical systems**) reviews Helmholtz's (1858) original contributions on topological aspects in fluid dynamics, presenting, then, a overview on current work on structural complexity analysis and some new results on topological bounds on magnetic energy of knots and links and helicity-crossing number relations. Finally, in **Random knotting: theorems, simulations and applications**, *De Witt Sumners* presents a very nice survey on random knotting and topological entanglement of filaments in space, starting from work on the Frisch-Wasserman-Delbruck Conjecture, and concluding with applications to viral DNA molecule packing and knotting. The volume is therefore a rather wide-ranging collection of important themes in current topological fluid mechanics and I can only hope that the interested reader may find here further stimulus to his/her own research work.

For all of this, I am particularly indebted, first of all to the authors, and secondly to the former Director of CIME, Arrigo Cellina, and to the present Director, Pietro Zecca, who patiently trusted in the aim of this project. I also want to thank Carla Dionisi for her technical support and last, but not least, Springer-Verlag, at Heidelberg.

University of Milano-Bicocca,   *Renzo L. Ricca*
March, 2008

Contents

Topological Quantities: Calculating Winding, Writhing,
Linking, and Higher Order Invariants 75
Mitchell A. Berger (CIME Lecturer)

Tangles, Rational Knots and DNA 99
Louis H. Kauffman (CIME Lecturer) and Sofia Lambropoulou

The Group and Hamiltonian Descriptions of Hydrodynamical
Systems 139
Boris Khesin (CIME Lecturer)

List of Contributors

Patrick D. Bangert
Algorithmica technologies GmbH
Ausser der Schleifmühle 67, 28203, Germany
e-mail: p.bangert@algorithmica-technologies.com
http://www.algorithmica-technologies.com

Mitchell A. Berger
Department of Mathematics, University of Exeter
North Park Road, Exeter, EX4 4QE, United Kingdom
e-mail: m.berger@exeter.ac.uk
http://secamlocal.ex.ac.uk/people/staff/mab215/home.htm

Louis H. Kauffman
Department of Mathematics, Statistics & Computer Science
University of Illinois, Chicago, 851 S. Morgan St.
Chicago IL 60607-7045, USA
e-mail: kauffman@math.uic.edu
http://www.math.uic.edu/~kauffman

Sofia Lambropoulou
Department of Mathematics National Technical University
Zografou Campus GR-15780 Athens, Greece
e-mail: sofia@math.ntua.gr
http://users.ntua.gr/sofial

Boris Khesin
Department of Mathematics, University of Toronto
100 St. George Street, Toronto, ON M5S 3G3, Canada
e-mail: khesin@math.toronto.edu
http://www.math.toronto.edu/khesin

H. Keith Moffatt
Department of Applied Mathematics & Theoretical Physics
Wiberforce Road, University of Cambridge, Cambridge
CB3 0WA, United Kingdom
e-mail: h.k.mofatt@damtp.cam.ac.uk

Renzo L. Ricca
Department of Mathematics and Applications, University of Milano-Bicocca
Via Cozzi 53, 20125 Milano, Italy
e-mail: renzo.ricca@unimib.it
http://www.matapp.unimib.it/~ricca/
and
Institute for Scientific Interchange, Villa Gualino, 10133 Torino, Italy

De Witt Sumners
Department of Mathematics, Florida State University, 208 Love Bldg.
Tallahassee, FL 320306-4510, USA
e-mail: sumners@math.fsu.edu

Braids and Knots

Patrick D. Bangert

Abstract We introduce braids via their historical roots and uses, make connections with knot theory and present the mathematical theory of braids through the braid group. Several basic mathematical properties of braids are explored and equivalence problems under several conditions defined and partly solved. The connection with knots is spelled out in detail and translation methods are presented. Finally a number of applications of braid theory are given. The presentation is pedagogical and principally aimed at interested readers from different fields of mathematics and natural science. The discussions are as self-contained as can be expected within the space limits and require very little previous mathematical knowledge. Literature references are given throughout to the original papers and to overview sources where more can be learned.

A short discussion of the topics presented follows. First, we give a historical overview of the origins of braid and knot theory (1). Topology as a whole is introduced (2.1) and we proceed to present braids in connection with knots (2.2), braids as topological objects (2.3), a group structure on braids (2.4) with several presentations (2.5) and two topological invariants arising from the braid group (2.6). Several properties of braids are then proven (2.7) and some algorithmic problems presented (2.8).

Braids in their connection with knots are discussed by first giving a notation for knots (3.1) and then illustrating how to turn a braid into a knot (3.2) for which an example is given (3.3). The problem of turning a knot into a braid is approached in two ways (3.4 and 3.5) and then the a complete invariant for knots is discussed (3.6) by means of an example.

P.D. Bangert
Algorithmica technologies GmbH
Ausser der Schleifmühle 67, 28203, Germany.
e-mail: p.bangert@algorithmica-technologies.com
http://www.algorithmica-technologies.com

R.L. Ricca (ed.), *Lectures on Topological Fluid Mechanics*,
Lecture Notes in Mathematics 1973, DOI: 10.1007/978-3-642-00837-5,
© Springer-Verlag Berlin Heidelberg 2009

The classification of knots is at the center of the theory. This problem can be approached via braid theory in several stages. The word problem is solved in two ways, Garside's original (4.1) and a novel method (4.2). Then the conjugacy problem is presented through Garside's original algorithm (4.3) and a new one (4.4). Markov's theorem allows this to be extended to classify knots but an algorithmic solution is still outstanding as will be discussed (4.5). Another important algorithmic problem, that of finding the shortest equal braid is presented at length (4.6).

We close with a list of interesting open problems (5).

Keywords: braids · knots · invariants · word problem · conjugacy problem · rewriting systems · fluid dynamics · path integration · quantum field theory · DNA · ideal knots

1 Physical Knots and Braids: A History and Overview

Possibly the most important difference between physical and mathematical knots is that mathematics requires the string to be closed. That means that after we tie the knot into a rope, we must glue the ends of the rope together and never undo them. The reason for this is that we are about to consider knots identical if we can continuously deform them into each other. If we had a rope with ends, we could untie the knot and thus every knot would be equal to a segment of straight rope.

The question of how braids are made is interesting in its own right. Suppose that we have n strings which are fixed at one end (we shall call this the *top* end) on a straight line and hang down vertically. The other ends are free to move in a horizontal plane P (the *bottom* end) below the top end. We further label each string by a number from one to n in order from left to right at the top end. Let the intersection of string i and plane P be labelled i also. We may now discuss the braid construction as a series of moves of the points 1 to n in the plane P relative to each other.

Let us begin with two points in P which interchange positions at every move, see figure 2 (a). This generates a braid on two strings which looks like figure 2 (b). The natural next step is to consider four points arranged in a square which interchange across the diagonals in turn, see figure 3 (a). This construction method is identical to taking the outer string of a four strand braid and passing it over two and under the string just overcrossed. We alternate between using the left and the right outer string for doing this.

Another way to generalise the scheme of figure 2 (a) is to introduce image points, see figure 3 (b). The image points differ from real points in that there are no strings attached to them. We exchange point and image point in numerical order. The way of constructing a braid shown in figure 3 (b) is one in which each string moves to a point which is its reflected image across

Fig. 1 Shells which display a clearly braided pattern (found by the author on the shore of Cetraro, Italy).

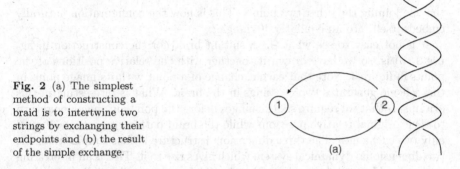

Fig. 2 (a) The simplest method of constructing a braid is to intertwine two strings by exchanging their endpoints and (b) the result of the simple exchange.

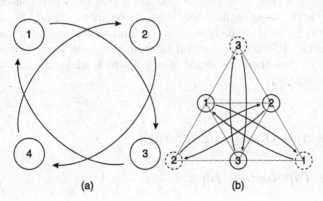

Fig. 3 (a) The points 1 and 3 and the points 2 and 4 form pairs which interchange their positions in turn, thus generating a braid and (b) the three points exchange positions with their image points (drawn in dashed circles) in turn.

Fig. 4 The braid which
results from the motion
described in figure 3 (b)
with the relative positions
of the points at the bottom
of each horizontal section.

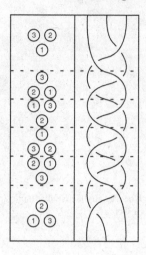

the line joining the other two points. This is how the configuration naturally embeds itself into an equilateral triangle.

It is not easy to see what the resultant braid for the construction in figure 3 (b) is, so we have drawn it together with the relative positions of the points in figure 4. Note that each exchange of a point with its image point in this scheme generates two crossings in the braid. What is interesting in this example is that we require six exchanges before the points in the plane return to their original relative positions while the braid pattern repeats itself after only two exchanges. The three dimensional structure is thus simpler than the two dimensional dynamical system which gives rise to it. This is an interesting property which can be exploited to classify fundamentally distinct motions in dynamical systems. It should be mentioned that the braid construction of figure 3 (b) is the most optimal way to stir a dye into a liquid [26]. The three points would be rods or paddles of some kind which would be submerged in the liquid and follow the motion prescribed. If one were to record their positions over time then one would obtain figure 4, where the time axis runs vertically upwards.

2 Braids and the Braid Group

2.1 The Topological Idea

Topology is a branch of mathematics that studies the shape of objects independent of their size or position. If we can deform one object into another by a continuous transformation, then we shall call these objects *topologically equivalent*. In everyday terms this means that we may bend, stretch, dent,

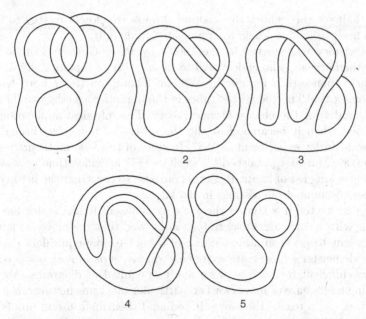

Fig. 5 This sequence of pictures shows how an initially linked structure (1) is slowly transformed into an unlinked structure (5) by a continuous transformation three stages of which are shown.

smoothen, move, blow up or deflate an object but we are not allowed to cut or to glue anything. An example of what we may do is shown in figure 5, in which it is shown by an explicit deformation that one can get from a linked structure to an unlinked structure.

Take for example the doughnut (also called the torus) and the sphere. These two objects are topologically inequivalent. This can be seen easily by observing that the doughnut has a hole while the sphere does not. We can only get rid of the hole by a discontinuous transformation, i.e. in the process of transforming the doughnut into the sphere there will be an instant at which the hole disappears. This is not allowed in topology and so we have motivated that there are topologically different objects.

2.2 The Origin of Braid Theory

Few areas of research can trace their origins as precisely as braid theory. Braid theory, as a mathematical discipline, began in 1925 when Emil Artin published his *Theorie der Zöpfe* [6]. A few problems in this first paper were quickly corrected [7] and the study was made algebraic soon thereafter [25].

As we shall see throughout this chapter, braids are closely related to knots and we need to look at knots to appreciate braids fully.

Knot theory was started in the 1860's by Peter Guthrie Tait, a Scottish mathematician, who endeavored to make a list of topologically distinct knots in response to a request by William Thompson (later Lord Kelvin) who thought that knotted vortex tubes in the luminiferous ether would make a good model for the elusive atomic theory. This physical application was abandoned when it became clear the the ether did not exist through the Michelson-Morley experiment in 1887. In spite of this, the mathematics was here to stay. Tait first published his work in 1877 at which time he was able to present a long list of knots. It was his purpose to construct the list in order of increasing number of crossings in the knot.

How can we tell if a knot is the same as another? Reidemeister has provided us with a convenient way to tell. He proved that the moves in figure 6 are sufficient to get from any diagram of a knot to any equivalent diagram [65]. Reidemeister's moves are extremely simple, it is almost obvious that they are sufficient to move between any two equivalent diagrams. Actually producing a sequence of moves for two particular diagrams however, is a nontrivial task. As a result, the moves have found their main use in proofs that certain quantities are identical for all diagrams of a particular knot. We call such quantities *invariants* and they can be integers, real numbers, polynomials, groups, manifolds and other mathematical objects. For example the number of components or closed loops of a knot is an integer invariant as the Reidemeister moves do not perform surgery (cutting or gluing).

Exercise 2.1. Convince yourself that any topological equivalence move possible on a knot can be reduced to Reidemeister moves.

There is only one knot of no crossings – a simple loop of rope – which is called the *unknot* (recall that a mathematical knot exists on rope without ends). No knot of one crossing can exist as this would simply be a twist of one end of the unknot and can just as easily be undone. This twist is the

Fig. 6 The Reidemeister moves.

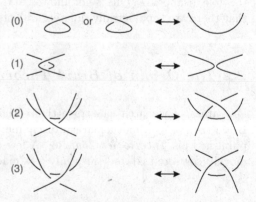

Fig. 7 The (a) trefoil knot
and (b) its mirror image.

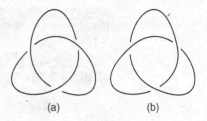

(a)   (b)

Reidemeister move zero in reverse, see figure 6. The same applies to any knot of two crossings. Matters become more challenging with three crossings because we hit the trefoil knot.

The trefoil knot (see figure 7 (a)) is the knot we usually tie into a shoelace before tying a bow on top of it. Is the trefoil knot the only knot of three crossings? We investigate this question by drawing three double points in the plane (do not differentiate between over and undercrossings yet) which are the intersections of two short line segments each. Then connect the endpoints of the line segments in all possible ways without causing further crossings in the plane. You will find that all of these will unravel to give the unknot by Reidemeister moves of type zero except the trefoil knot. The trefoil knot comes in two natural flavours: the standard type (figure 7 (a)) and the mirror image of the standard type (figure 7 (b)). The *mirror image* of a knot is obtained by switching all of its crossings (see figure 7 (b)). This method is essentially the one which Rev. Kirkman used in the 19th century to construct a list of all knots up to and including ten crossings. The labour involved in this task is prodigious. To complete the list of all knots of three crossings we must ask:

Exercise 2.2. Is the trefoil equivalent to its mirror image? [Hint: This means that you have to find a continuous deformation of the trefoil into its mirror image if they are equivalent or a proof that it is not possible otherwise. The typical method is to look for an invariant if you suspect that they are not equivalent. If you can show that the value of the invariant is different for the two knots, then you have shown that they are different.]

The answer to exercise 2.2 is that the trefoil is *not* equivalent to its mirror image. This is shown by computing an invariant quantity called *Alexander polynomial* for both knots. We will compute the polynomials for both trefoils in section 2.5.

We will motivate the result here by computing a quantity called *writhe* which is an invariant of all the Reidemeister moves except the zeroth one. Imagine you want to hang a painting on the wall and you are putting a screw into the wall to hold the painting up. You twist the screw clockwise to get it into the wall and counterclockwise when you've made a mistake and wish to get it out again. We consider progress positive and mistakes negative so that a crossing in a knot which is achieved by a clockwise rotation of the hands

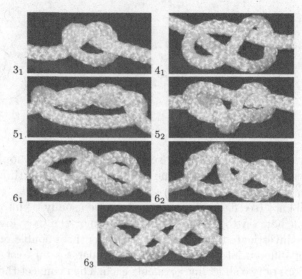

Fig. 8 The prime knots with fewer than seven crossings and their names from standard tables. The knot 3_1 is the trefoil knot of figure 7 (a).

as they follow the orientation of the knot is assigned a weighting of $+1$; the opposite kind is assigned a weight of -1. The writhe w of a knot is the sum of the weights over all the crossings. Let us pick the orientation in which the topmost arch on the trefoils in figure 7 points to the left. Then the standard trefoil has $w = 3$ and the mirror image $w = -3$.

Exercise 2.3. Prove that writhe is invariant for all Reidemeister moves except move zero.

Using such methods, it is possible to construct a large table of knots. In figure 8 we show the first seven knots after the unknot. It is understood that the two ends of the rope must be joined to yield the mathematical knot. We present them in this fashion for ease of understanding and practical experimentation.

The question is: Can we find a general method to determine equality or otherwise for any two knots? The answer is yes, but with qualifications. There exists a method due to Waldhausen, Hakken, Hemion and others but it is so inefficient that it is not possible to use for knots for which we do not know the answer already [39]. There exists a theorem due to Alexander which states that every knot can be represented by a braid [4] which we prove in theorem 2.9. This gave the motivation for people to study braids in order to try to help classify knots. The greatest thrust came from Markov who proved a result for braids similar to Reidemeister move result for knots [55]. Using Markov's theorem to classify knots has proven difficult however and the search continues.

Braids have proven tremendously useful in spite of the fact that they have not led to a complete knot classification scheme. Many invariant of knots are naturally defined on braids. The most revolutionary invariant, the Jones polynomial, was discovered using braid theory. Beyond this, braids have many applications to various fields as we shall discover in the sections to come.

An operation to combine knots can be defined which we are going to call knot addition and denote it by #.

Definition 2.4. Given two knots K and L, we define the *knot sum* $K\#L$ as the knot obtained by cutting both K and L at a random location and gluing them together with respect to their orientations.

This is a simple operation but it is not at all obvious that it is well defined. One can show that: (1) The sum is independent of the points on K and L chosen as cutting points [60], (2) any knot can be uniquely factorized into a finite length sum of knots [67], (3) this sum may actually be determined [68]. Property (1) makes the concept well-defined. The second property establishes the existence of *prime knots*, i.e. knots that may not be decomposed into the sum of others and also that classifying prime knots will classify all knots. The third property means that this is, at least in theory, possible to actually compute. However, the algorithm to find the unique decomposition is the algorithm alluded to previously and thus this is not a practical method. It can be shown that there does not, in general, exist an inverse to the operation of addition of knots. As one may show that knot addition is associative, knots form a semi-group but not a group under the operation of addition [60].

2.3 The Topological Braid

We know from section 1 what a braid is. Mathematically, we have to be slightly more careful.

Definition 2.5 (n-braid). Let l_1 and l_2 be two parallel lines in a plane P and let $A = \{a_1, a_2, \cdots, a_n\}$ and $B = \{b_1, b_2, \cdots, b_n\}$ be sets of points on l_1 and l_2 respectively. An *n-braid* is a set of (possibly oriented) n non-intersecting polygonal curves which have exactly one endpoint in A and one in B such that all points in A or B are the endpoints of exactly one of these curves and such that any line l parallel to l_1 and l_2 crosses any curve in at most one point.

An example of a 3-braid, in which we have labelled the lines l_1 and l_2 as well as the point sets A and B, is shown in figure 9 (a); note that the plane P is understood to be the plane of the paper. In this example, we

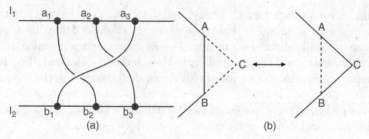

Fig. 9 (a) An example of the definition of a topological braid (see definition 2.5) and (b) an elementary deformation as defined in definition 2.6.

have 3 curves each of which have two endpoints, one in A and one in B. These curves go from l_1 to l_2 monotonically, they do not double back on themselves. This is the meaning that no line parallel to l_1 and l_2 may cross any curve in more than one point. In fact it is this requirement that makes braids substantially simpler than knots and allows a group structure to be defined on braids. An n-braid in the form of definition 2.5 is also called an *open braid*. We shall drop the n from n-braid when no confusion can arise.

The normal braid which is braided into peoples' hair fulfills these requirements. One end of all the strings is fixed on the person's head and others are held in place by some form of rubber band. The braiding in the middle is done in a way that each bundle of hairs goes from top to bottom monotonically. Some Celtic designs used as borders in the Book of Kells or other illuminated books or more commonly used as trimmings for medieval clothing, necklaces, pendants and belts are not usually braids conforming to this definition as their strings often return to a point close to their origin and thus contain a local maximum or minimum.

Whenever we define a new mathematical object, we desire an equivalence relation for the possible instances of this object. As braid theory is a topological pursuit, we will allow ourselves the usual freedom of topology which means that we will allow the object to be distorted in any way as long as this can be done continuously. So we may bend, stretch and pull a string but we may never cut a string or glue two strings together; such actions are called *surgery* and are said to change the topology. For braids, it is clear that if we do not fix the endpoints of the curves, we shall be able to transform any braid into any other yielding a rather boring theory; thus we also require the ends to be fixed. We will call the equivalence relation for braids under these conditions *isotopy*. Before we can define isotopy, we must define what we mean by a topological deformation. An elementary deformation is the basis for all topological deformations.

Definition 2.6 (elementary deformation). Suppose that a braid string (recall that it was defined as a polygonal curve) has points A and B as vertices. We may then create a further point C, delete the segment AB and create the segments AC and CB. This deformation, and its inverse, is called an *elementary deformation* if and only if the triangle ABC does not intersect any other strings and only meets the current string along its side AB. See figure 9 (b) for an illustration.

Definition 2.7 (braid isotopy). Two braids α and β are called *isotopic*, denoted $\alpha \approx \beta$, if and only if α can be transformed into β using a finite number of elementary deformations.

Suppose we were to label the string which intersects the point b_i by i. On l_2, the string labels from left to right would thus be in numerical order whereas on l_1 they may not be ordered numerically. If we list the numerical labels of the strings which intersect l_1 from left to right, we obtain a permutation on the set of integers $\{1, 2, \cdots, n\}$. A braid thus induces a permutation on the set of the first n integers. For example, the braid in figure 9 induces the permutation $[2, 3, 1]$. The fact that the induced permutation is an equivalence class invariant follows immediately from the requirement that the endpoints be fixed.

An open braid may be *closed* to yield a knot. See figure 10 for the closure of the braid from figure 9. A *closed braid*, denoted $\overline{\alpha}$, is obtained from an open n-braid α by deleting l_1 and l_2 and connecting points a_i and b_i with non-intersecting polygonal curves in P for all $i : 1 \leq i \leq n$. It is clear that the closure of any braid yields a knot. Thus some knots may be represented as closed braids. Unfortunately determining the closed braid, given a knot, is not so easy. It is however possible and we shall solve this problem in sections 3.4 and 3.5. The proof of the fact that *all* knots may be represented as closed braids, Alexander's theorem, gave the initial momentum for studying braids in detail [4]. Artin took up the challenge and constructed a theory of braids with a view to use them to deal with knots.

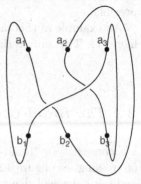

Fig. 10 The braid from figure 9 is closed here. It is immediate upon simple transformation that this knot is the same as the simple loop or the unknot.

Exercise 2.8. A simple knot invariant is the number of components a knot has. Show that the number of components of the knot $\overline{\alpha}$ is equal to the number of cycles in the permutation that the braid α induces.

Alexander's theorem is usually proved by giving a topological method with which to deform a knot into a closed braid. There exist several distinct methods of doing so but most are not suitable for use; they are only employed to establish the theorem. The proof given here is fundamentally different and new to the best knowledge of the author. The theorem assumes that the knot is oriented but if not the transformation is accurate up to orientation change.

Theorem 2.9 (Alexander [4]). *Every knot may be represented as a closed braid.*

Proof. Consider an oriented straight line a in $\mathbb{R}^3$ which we will call the axis. Choose a point O on a and construct a cylindrical polar coordinate system which has O as its origin. The positive z, or upward vertical, direction is directed parallel to a in the direction of its orientation. The polar angle ϕ increases in the counterclockwise direction, as usual. Using this system, the theorem claims that every knot K can be deformed with respect to a in such a way that the polar angle of a point P going along any component of K strictly increases or more simply: As we travel along the knot we will go around the axis a without ever changing our counterclockwise direction. Suppose we have n straight line segments s_i for $1 \leq i \leq n$ with endpoints R_i and S_i such that the polar angle of S_i, $\phi(S_i)$ is is larger than $\phi(R_i)$. We may form any knot by subdividing these segments into a finite number of straight subsegments, moving the endpoints of the subsegments and performing surgery which identifies R_i and S_i for all i. Here, we will form the knot K by keeping R_i fixed and moving the point S_i creating new points $Q_{i,j}$ indexed by j as necessary. Whenever it becomes necessary to move S_i to a position of lower polar angle than the last $Q_{i,j}$ created, move S_i once around a creating a suitable number of points doing so and then continuing. After the required knot is formed, we perform the surgery of identifying R_i and S_i. By definition of a knot as a polygonal curve such a construction is always possible.

An example of this method applied to forming the trefoil knot is given in figure 11. This proves the theorem.   □

2.4 The Braid Group

We note that any braid can be represented by a vertical stack of two types of crossing, see figure 12. When all strings are vertical apart from strings i and $i + 1$, we will denote this crossing by σ_i or σ_i^{-1} depending on whether

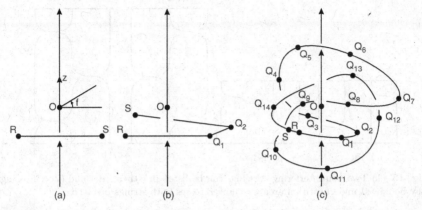

Fig. 11 Constructing the trefoil knot as a closed braid, see proof of theorem 2.9 for a discussion.

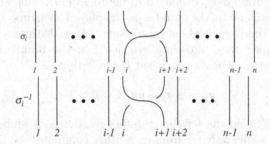

Fig. 12 The generator σ_i and its inverse σ_i^{-1} for the braid group B_n.

string i overcrosses or undercrosses string $i+1$ respectively. It is thus clear that any braid can be specified by a string of these symbols. We agree to the convention that the left to right direction of the symbols representing a braid shall correspond to the upward direction of the braid; that is, the lowest crossing corresponds to the first symbol. This is a convention and some other authors use the opposite convention. While care is required, no serious consequences arise from this choice. For example, the braid in figure 2 (b) is σ_1^5 (the power means that the symbol σ_1 was repeated five times) and the braid in figure 4 is $\left(\sigma_1\sigma_2^{-1}\sigma_1^{-1}\sigma_2\right)^3$. From now on, we shall denote a braid by these symbols.

Definition 2.10 (braid word). We will call any sequence of $\sigma_i^{\pm1}$ a *braid word*.

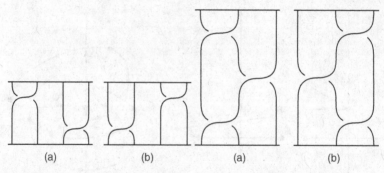

(a) (b) (a) (b)

Fig. 13 (1) Two non-interfering crossings can be listed in either order and (2) a crossing may be moved underneath an arch which overcrosses both strings involved

Definition 2.11 (positive and negative braid word). If a braid word contains only σ_i (and no σ_i^{-1}) then it will be called *positive*. However, if it is contains only σ_i^{-1} (and no σ_i) then it will be called *negative*.

Consider the braid $\sigma_3\sigma_1$ displayed in figure 13 (1a). This braid is clearly topological equivalent to the braid $\sigma_1\sigma_3$ displayed in figure 13 (1b). More generally, every time two neighboring crossings are on distinct pairs of strings, the order in which these crossings are listed in the braid word does not topologically matter. Thus we arrive at the rule that

$$\sigma_i\sigma_j \approx \sigma_j\sigma_i \text{ for } |i - j| > 1 \tag{1}$$

which is usually called the *far commutation* relation as it embodies the fact that generators sufficiently far from each other commute.

It also becomes clear that if we have an arch which over or undercrosses two strings which then cross (see figure 13 (2a)), this crossing may be moved onto the other side of the arch (see figure 13 (2b)). Thus the braids $\sigma_1\sigma_2\sigma_1$ and $\sigma_2\sigma_1\sigma_2$ are topologically equivalent. As this can hold anywhere in a braid, we arrive at the second rule that

$$\sigma_i\sigma_{i+1}\sigma_i \approx \sigma_{i+1}\sigma_i\sigma_{i+1} \tag{2}$$

which is typically called the *braid relation*. We find this name too vague and so we will refer to this relation as the *bridge relation* as it symbolizes that anything not in conflict with the principal pillars may move freely both below and above a bridge.

After some experimentation, one notices that all the moves one may make on a braid while preserving its topology can be reduced to applying the rules in equations 1 and 2 to the braid word. We would like to prove that this is always so.

Theorem 2.12. *(Artin [6] [7]) The equivalence relation upon braid words defined by the relations 1 and 2 is identical to the equivalence relation of braid isotopy (see definition 2.7) upon the braids represented by the braid words.*

Proof. Recall that any knot may be represented as a closed braid. The braid contains all the crossings and Reidemeister's moves define equivalence of knots. Thus braid equivalence is Reidemeister equivalence of the braid. Move zero would create an object which is not a braid but the others apply. Translating these into the σ_i notation and simplifying yields the given presentation. □

From figure 12, it is clear that σ_i^{-1} is the inverse of σ_i. As we represent a braid by a braid word in the σ_i from the bottom up, we can easily concatenate two braids together. The braid $\alpha\beta$ is constructed from the braids α and β by identifying the top ends of α with the bottom ends of β. It is obvious that concatenation is associative, i.e. $(\alpha\beta)\gamma \approx \alpha(\beta\gamma)$ and that the concatenation of two braids is a braid. Suppose ι is the braid of n vertical strings and no crossings. We have $\iota\alpha \approx \alpha\iota \approx \alpha$ for any n-braid α. The braid ι acts as an identity. Since we have closure, associativity, inverses and an identity, the set of n-braids forms a group generated by the generators σ_i. We will denote this group by B_n and refer to this family of groups as the *braid groups*. Because of theorem 2.12, B_n has the presentation

$$B_n = \left\langle \{\sigma_1, \sigma_2, \cdots, \sigma_{n-1}\} : \begin{array}{l} \sigma_i\sigma_{i+1}\sigma_i \approx \sigma_{i+1}\sigma_i\sigma_{i+1}, \\ \sigma_i\sigma_j \approx \sigma_j\sigma_i \text{ for } |i-j| > 1 \end{array} \right\rangle \qquad (3)$$

It is instructive to consider the braid group from another point of view which curiously leads to the same presentation as given above in equation 3. Consider the space between the two parallel planes which contain l_1 and l_2 respectively after the braid has been removed from it. This space has a fundamental group which may be represented by a series of loops beginning and ending on some randomly chosen base point b and going around the removed braid strings. There are n distinct such loops which we shall call x_i with i running from 1 to n. The fundamental group is free of rank n with the x_i as generators. x_1 is the loop around the first string from the left on each level and so on for the other x_i. From level to level a reassignment of generators becomes necessary. This is called an automorphism and the particular one we need here, a_i is defined by

$$a_i : x_i \to x_{i+1}; \ x_{i+1} \to x_{i+1}^{-1}x_ix_{i+1}; \ x_p \to x_p \ (p \neq i, i+1) \qquad (4)$$

The map $\alpha : \sigma_i \to a_i$ is a homomorphism of B_n into the automorphism group of F_n, the free group of rank n. It can be shown that the a_i generate a group with presentation identical to equation 3 under the homomorphism α. In other words, a braid word may be regarded as an automorphism of F_n. As the mapping merely consists of a change of symbol for the generators, it is frequently useful not to make a distinction between a braid word as an element of B_n and as an automorphism of F_n. In the next section, we will introduce some other presentations of the braid group.

2.5 Other Presentations of the Braid Group

The presentation of B_n given in equation 3 is called the *Artin presentation* as it was Artin who first used it in the paper which founded the field. The fact that braids admit a group structure simplifies their treatment tremendously. It can be shown that knots do not admit a group structure and this is one reason why the problem of deciding if two knots are equal is so different from the similar question about braids. It is difficult, however to extract useful information from a presentation of a group. For this reason it is useful to search for other presentations of the same group with special properties.

The Artin presentation has the appeal that it is very topological. It is easy to draw the braid given the braid word and it is easy to read off the braid word from a braid. Furthermore, both the far commutation and the bridge relations are simple to perform. One disadvantage is that each braid group has a different number of generators. Consider putting $a = \sigma_1\sigma_2\cdots\sigma_{n-1}$ and $\sigma = \sigma_1$. After some manipulation, we find the presentation

$$B_n = \left\langle \{a,\sigma\} : a^n \approx (a\sigma)^{n-1}, \ \sigma a^{-j}\sigma a^j \approx a^{-j}\sigma a^j\sigma \text{ for } 2 \leq j \leq \frac{n}{2} \right\rangle \quad (5)$$

which we call the *Coxeter presentation*. The Coxeter presentation has the advantage that all braid groups have just two generators but we have lost some of the topological correspondence.

It would be nice to have a matrix representation of the braid groups. To this end, we will represent the identity matrix of n rows and columns by I_n (recall that the identity matrix has unity entries in the leading diagonal and zeros everywhere else). Then we define the mapping $\phi_n(\sigma_i)$ of an Artin generator σ_i to a matrix of n rows and n columns whose entries are Laurent polynomials in the variable t (Laurent polynomials allow both positive and negative powers of the variable). Expressed formally, this means

$$\phi_n : B_n \to GL\left(n, \mathbf{Z}\left[t^{\pm 1}\right]\right) \quad (6)$$

where $\mathbf{Z}$ is the ring of polynomials. The mapping is defined by

$$\phi_n(\sigma_i) = \begin{pmatrix} I_{i-1} & 0 & 0 & 0 \\ 0 & 1-t & t & 0 \\ 0 & 1 & 0 & 0 \\ 0 & 0 & 0 & I_{n-i-1} \end{pmatrix} \quad (7)$$

It can easily be shown that ϕ_n is a homomorphism, i.e. that $\phi_n(\sigma_i\sigma_j) = \phi_n(\sigma_i)\phi_n(\sigma_j)$ where multiplication in $GL\left(n, \mathbf{Z}\left[t^{\pm 1}\right]\right)$ is the usual matrix multiplication. A representation is termed *faithful* when the mapping ϕ giving rise to it is injective, i.e. when $x \neq y$ implies $\phi(x) \neq \phi(y)$. This representation is faithful for $n \leq 3$ [54] and not for $n \geq 5$ [13]. For $n = 4$ the answer is unknown.

It should now be easy to write down a matrix representation for any braid word. For example, if $n = 3$, then for $\phi_n(\sigma_1\sigma_2)$ we have

$$\phi_n(\sigma_1)\phi_n(\sigma_2) = \begin{pmatrix} 1-t & t & 0 \\ 1 & 0 & 0 \\ 0 & 0 & 1 \end{pmatrix} \begin{pmatrix} 1 & 0 & 0 \\ 0 & 1-t & t \\ 0 & 1 & 0 \end{pmatrix} = \begin{pmatrix} 1-t & t-t^2 & t^2 \\ 1 & 0 & 0 \\ 0 & 1 & 0 \end{pmatrix} \quad (8)$$

This representation is called the *Burau representation*.

Recently a new presentation was invented by Birman, Ko and Lee [15] which they used to solve the word and conjugacy problem in a new way. The generators a_{kl} are defined by

$$a_{kl} = (\sigma_{k-1}\sigma_{k-2}\cdots\sigma_{l+1})\,\sigma_l\,\left(\sigma_{l+1}^{-1}\sigma_{l+2}^{-1}\cdots\sigma_{k-1}^{-1}\right) \quad (9)$$

Topologically this is a crossing between two arbitrary braid strings k and l. In a_{kl}, string k overcrosses string l and both strings overcross all other string in between them to be able to cross and then overcross the in between strings again to return to their original (but now switched) positions. Using these generators, B_n has the presentation

$$B_n = \left\langle \{a_{ts}; n \geq t > s \geq 1\} : \begin{array}{c} a_{ts}a_{sr} = a_{tr}a_{ts} = a_{sr}a_{tr}, \\ a_{ts}a_{rq} = a_{rq}a_{ts} \text{ for} \\ (t-r)(t-q)(s-r)(s-q) > 0 \end{array} \right\rangle \quad (10)$$

which we call the *band-generator presentation*. In the Artin representation, we number the strings from left to right on each level. Suppose we were to label each string with a unique label which it would carry throughout the braid. At the bottom of the braid, we number the strings from one to n as we go from left to right. Each crossing in which string i overcrosses string j is labelled with the generator g_{ij}. This presentation is called the *colored braid presentation* as the string labels act like each string was made of a separate color. This representation retains the complete information of the braid but it is not immediately apparent whether a crossing is positive or negative in the Artin sense. Note that this feature means that the generators g_{ij} are self inverse, $g_{ij}^2 \approx e$ the identity. We easily write down the braid group relations in this presentation to get

$$B_n = \left\langle \{g_{ij}\} \text{ for } \begin{array}{c} 1 \leq i, j \leq n \\ i \neq j \end{array} : \begin{array}{c} g_{ij}g_{ik}g_{jk} = g_{jk}g_{ik}g_{ij}, \\ g_{ij}g_{kl} = g_{kl}g_{ij} \text{ for} \\ (i-k)(i-l)(j-k)(j-l) > 0 \end{array} \right\rangle \quad (11)$$

We note that in the colored braid presentation, the fundamental braid Δ_n takes the form

$$\Delta_n = g_{12}g_{13}\cdots g_{1n}g_{23}g_{24}\cdots g_{n-1\,n} \tag{12}$$

$$= \prod_{i=1}^{n-1}\prod_{j=i+1}^{n} g_{ij} \tag{13}$$

2.6 The Alexander and Jones Polynomials

The Burau representation of the braid group is important in the definition of a revolutionary knot-invariant called the *Alexander polynomial*. We begin with a braid word α, construct its Burau representation $\phi_n(\alpha)$ and take the determinant of the matrix $[\phi_n(\alpha) - I_n]_{1,1}$ where the subscript indicates that the first row and column should be deleted. This determinant can be shown to be a topological invariant of the closure of the braid α and is denoted by $\Delta_{\overline{\alpha}}$ such that

$$\Delta_{\overline{\alpha}}(t) = \det[\phi_n(\alpha) - I_n]_{1,1} \tag{14}$$

Thus the Alexander polynomial of the braid $\sigma_1\sigma_2$ of the above example is $\Delta = 1$ which happens to be the same value as the Alexander polynomial for the unknot. While $\overline{\sigma_1\sigma_2}$ is actually isotopic to the unknot, we could not conclude this from its Alexander polynomial. The Alexander polynomial is an *incomplete* invariant in that we can only say that if $\Delta_K \neq \Delta_{K'}$, then $K \neq K'$. The converse is not true, in general. Nevertheless, the Alexander polynomial is very important in knot theory. Let us compute the Alexander polynomial for both trefoils (see figure 7). The ordinary trefoil is the closure of σ_1^3 and its mirror is the closure of σ_1^{-3}. Now we have

$$\phi_2\left(\sigma_1^3\right) = \begin{pmatrix} (1-t)^3 & 0 \\ 0 & t^3 \end{pmatrix} \qquad \phi_2\left(\sigma_1^{-3}\right) = \begin{pmatrix} (1-t)^{-3} & 0 \\ 0 & t^{-3} \end{pmatrix} \tag{15}$$

and thus

$$\Delta_{\overline{\sigma_1^3}} = t^3 - 1 \qquad \Delta_{\overline{\sigma_1^3}} = t^{-3} - 1 \tag{16}$$

And since the Alexander polynomials of the two trefoils are distinct, the knots must be distinct.

Many other polynomials invariants have been devised after the Alexander polynomial, the most important is the Jones polynomial. The Jones polynomial is also an incomplete invariant of knots which was originally constructed in terms of braids. We shall not prove that it is an invariant, nor that it is incomplete but we shall give an easy method to obtain it. Even though it is incomplete, it is a very powerful invariant in that it distinguishes many knots not distinguished by other invariants such as the Alexander polynomial.

We define the *tensor product* of a $p \times q$ matrix A and a $r \times s$ matrix B by

$$A \otimes B = \begin{pmatrix} a_{11}B & a_{12}B & \cdots & a_{1q}B \\ a_{21}B & a_{22}B & \cdots & a_{2q}B \\ \vdots & \vdots & \ddots & \vdots \\ a_{p1}B & a_{p2}B & \cdots & a_{pq}B \end{pmatrix} \tag{17}$$

$A \otimes B$ is clearly a $pr \times qs$ matrix. We also define the *trace* of a $n \times n$ matrix C by

$$tr(C) = \sum_{i=1}^{n} c_{ii} \tag{18}$$

Consider the matrix

$$\mu(t) = \begin{pmatrix} 1 & 0 \\ 0 & t \end{pmatrix} \tag{19}$$

and define

$$\mu^{\otimes n} = \underbrace{\mu \otimes \mu \otimes \cdots \otimes \mu}_{n \text{ times}} \tag{20}$$

then $\mu^{\otimes n}$ is a $2^n \times 2^n$ matrix and $tr\left(\mu^{\otimes n}\right) = (1+t)^n$. Furthermore let

$$R = \begin{pmatrix} 1 & 0 & 0 & 0 \\ 0 & 0 & -t^{1/2} & 0 \\ 0 & -t^{1/2} & 1-t & 0 \\ 0 & 0 & 0 & 1 \end{pmatrix}; \quad R^{-1} = \begin{pmatrix} 1 & 0 & 0 & 0 \\ 0 & 1-t^{-1} & -t^{-1/2} & 0 \\ 0 & -t^{-1/2} & 0 & 0 \\ 0 & 0 & 0 & 1 \end{pmatrix} \tag{21}$$

where we use the convention that $\left(t^{1/2}\right)^2 = t$.

After all these definitions, we are ready to define a map from the braid group B_n to the $2^n \times 2^n$ matrices with Laurent polynomial entries ($\Phi_n : B_n \to M\left(2^n, \mathbb{Z}\left[t^{1/2}, t^{-1/2}\right]\right)$) given by

$$\Phi_n\left(\sigma_i^{\epsilon}\right) = \underbrace{I_2 \otimes I_2 \otimes \cdots I_2}_{i-1 \text{ times}} \otimes R^{\epsilon} \otimes \underbrace{I_2 \otimes I_2 \otimes \cdots I_2}_{n-i-1 \text{ times}} \tag{22}$$

where $\epsilon = \pm 1$ and I_2 is the 2×2 identity matrix. For some general braid word β we have

$$\beta = \sigma_{i_1}^{\epsilon_1} \sigma_{i_2}^{\epsilon_2} \cdots \sigma_{i_k}^{\epsilon_k} \quad (1 \le i_1, i_2, \cdots, i_n < n; \; \epsilon_j = \pm 1) \tag{23}$$

$$\Phi_n(\beta) = \Phi_n\left(\sigma_{i_1}^{\epsilon_1}\right) \Phi_n\left(\sigma_{i_2}^{\epsilon_2}\right) \cdots \Phi_n\left(\sigma_{i_k}^{\epsilon_k}\right) \tag{24}$$

If we now define

$$\epsilon(\beta) = \epsilon_1 + \epsilon_2 + \cdots + \epsilon_k \tag{25}$$

then the Jones polynomial is given by

$$V_{\overline{\beta}}(t) = \frac{t^{\frac{e(\beta)-n+1}{2}} tr\left(\Phi_n(\beta)\mu^{\otimes n}\right)}{1+t} \tag{26}$$

and we have that if $\overline{\beta_1} \approx \overline{\beta_2}$, then $V_{\overline{\beta_1}}(t) = V_{\overline{\beta_2}}(t)$ for any two braids β_1 and β_2 (not necessarily members of the same braid group).

Exercise 2.13. Compute the Jones polynomial for the 2-braid σ_1, the closure of which is clearly the unknot (solution follows).

We just compute and obtain

$$V_{\overline{\sigma_1}}(t) = \frac{t^{\frac{1-2+1}{2}}}{1+t} tr\left[\begin{pmatrix} 1 & 0 & 0 & 0 \\ 0 & 0 & -\sqrt{t} & 0 \\ 0 & -\sqrt{t} & 1-t & 0 \\ 0 & 0 & 0 & 1 \end{pmatrix}\begin{pmatrix} 1 & 0 & 0 & 0 \\ 0 & t & 0 & 0 \\ 0 & 0 & t^2 & 0 \\ 0 & 0 & 0 & t^3 \end{pmatrix}\right] = 1 \tag{27}$$

It is a well-known conjecture that the Jones polynomial recognizes the unknot, that is to say that the Jones polynomial is equal to 1 if and only if the knot is isotopic to the unknot. Many people have been searching for a proof and a counterexample and none has been found thus far. Resolving this question is one of the more important open problems in knot theory.

Proving that the Jones polynomials is an invariant of closures of braids is easy. For reasons of space, we simply outline the equations that need to be checked by straightforward calculations.

Theorem 2.14. *The Jones polynomial is a knot invariant.*

Proof. By direct calculation, check

1. The mapping Φ_n is a homomorphism from B_n to $M\left(2^n, \mathbb{Z}\left[t^{1/2}, t^{-1/2}\right]\right)$. This can be done by checking that

$$\Phi_n(\sigma_i\sigma_j) = \Phi_n(\sigma_j\sigma_i); \quad \Phi_n(\sigma_i\sigma_{i+1}\sigma_i) = \Phi_n(\sigma_{i+1}\sigma_i\sigma_{i+1}) \quad (|i-j| > 1) \tag{28}$$

2. By Markov's theorem, two closed braids are isotopic if and only if they can be reached from each other by conjugacy and stabilization. Thus we need to check if the Jones polynomial is invariant under these two moves,

$$V_{\overline{\gamma\beta\gamma^{-1}}}(t) = V_{\overline{\beta}}(t); \quad V_{\overline{\beta}}(t) = V_{\overline{\beta\sigma_n^{\pm 1}}}(t) \quad (\gamma, \beta \in B_n) \tag{29}$$
$$\square$$

The polynomial invariants can be calculated using so-called *skein relations*. These relations are expressions relating the polynomials of knots that are identical except for a single local change. In braid language, the Alexander polynomial of the closed braid $\beta \in B_n$, $A_\beta(z)$ is given by the relation

$$A_{\beta\sigma_i}(z) - A_{\beta\sigma_i^{-1}}(z) = zA_\beta(z) \tag{30}$$

together with the boundary condition that if the closure of β is the unknot, $A(z) = 1$. The Jones polynomial $J_\beta(z)$ has the same boundary condition but takes the skein relation

$$\frac{J_{\beta\sigma_i}(z)}{z} - tJ_{\beta\sigma_i^{-1}}(z) = \left(\sqrt{z} - \frac{1}{\sqrt{t}}\right)J_\beta(z) \tag{31}$$

Using these skein relations, it is possible to calculate the polynomial invariants straight from the diagram by successively eliminating crossings [48].

2.7 Properties of the Braid Group

In this section we will prove various propositions which enumerate the basic properties of braids. The Artin presentation will be used for all these as for most of the rest of the chapter.

Recall that the center of a group is the set of all those elements which commute with all other elements in that group. The centre of the braid groups will become important in many places and so we will construct it.

Definition 2.15 (fundamental word). The *fundamental braid word* $\Delta_n \in B_n$ is defined by

$$\Delta_n = \sigma_1\sigma_2\cdots\sigma_{n-1}\sigma_1\sigma_2\cdots\sigma_{n-2}\cdots\sigma_1\sigma_2\sigma_1 \tag{32}$$

It is a simple matter of applying the braid group relations to find that

$$\Delta_n^2 = (\sigma_1\sigma_2\cdots\sigma_{n-1})^n \tag{33}$$

Proposition 2.16. *The center $C(B_n)$ of the braid group B_n is the set*

$$C(B_n) = \left\{\Delta_n^{2i}\right\} \tag{34}$$

for any (positive, negative or zero) integer i.

Definition 2.17. The *ascending braid word* a_j is defined by $a_j = \sigma_1\sigma_2\cdots\sigma_j$.

Proposition 2.18. *In B_n, we have $\sigma_i a_j \approx a_j \sigma_{i-1}$ for $1 < i \le j < n$.*

Proposition 2.19. *For all i we have $\sigma_i\Delta_n \approx \Delta_n\sigma_{n-i}$.*

Definition 2.20. Two positive n-braid words a and b are called *positively equal* if and only if there exists a sequence of words W_i with $0 \le i \le p$ for some finite p for which $W_0 = a$, $W_p = b$, W_i is obtained from W_{i-1} by a single application of one of the defining relations of B_n, and all W_i are positive.

In other words, two braids are positively equal if we can transform one into the other without ever having to use an inverse generator.

Proposition 2.21. *Two equal positive braid words are positively equal.*

2.8 Algorithmic Problems in the Braid Groups

We have defined braids, elucidated their connection with knots and found a group structure on braids. This group structure has certain properties of which we enumerated a few important ones in the last section. There are a number of questions which we may readily ask about braids, the most significant of which we shall describe in this section. Clearly, we wish to know both necessary and sufficient conditions for equivalence.

Definition 2.22 (word problem). Given two braid words $\alpha, \beta \in B_n$, the *word problem* asks whether $\alpha \approx \beta$.

The question of whether $\alpha \approx \beta$ is identical to the question of whether $\alpha\beta^{-1} \approx e$, the identity in B_n. Thus the word problem reduces to recognizing the identity element.

Recall that two elements a and b of some group G are called *conjugate*, denoted $a \approx_c b$, if and only if there exists a $c \in G$ such that $a \approx cbc^{-1}$. The conjugacy condition becomes $\alpha\gamma \approx \gamma\beta$ for two braid words $\alpha, \beta \in B_n$ to be conjugate with respect to a third braid word $\gamma \in B_n$. We are naturally interested under what conditions such a commutation relation exists.

Definition 2.23 (conjugacy problem). Given two braid words $\alpha, \beta \in B_n$, the *conjugacy problem* asks whether there exists a third braid word $\gamma \in B_n$ such that $\alpha\gamma \approx \gamma\beta$.

Suppose $\alpha \approx \beta$, then we also have $\alpha\gamma \approx \gamma\beta$ for $\gamma \approx e$. Thus any equal braid words are also conjugate but two conjugate braid words are not necessarily equal and so the conjugacy problem subsumes the word problem.

The most central question in knot theory is that of classification: Given two knots K_1, K_2 are they topologically equivalent $K_1 \approx K_2$? Markov has found it possible to translate this question into a question about braids. Imagine closing the braid cbc^{-1}. We may move the subbraid c through the closure part so that it emerges on top of the rest of the braid, i.e. becomes $bc^{-1}c \approx b$. In other words, conjugation of a braid preserves closed braid isotopy: Any two conjugate braids represent equivalent knots. Conjugation is a sufficient condition but unfortunately not necessary. The braid $e \in B_1$ is just a single vertical line segment which closes to the unknot. The braid $\sigma_1 \in B_2$ is the braid of single positive crossings which also closes to the unknot. As the braids are in different braid groups, they have a different number of strings and are thus not conjugate to each other. Such braids can be related through a move called stabilization.

Definition 2.24 (stabilization move). Given a braid $\alpha \in B_n$, the operation $\alpha \leftrightarrow \alpha\sigma_n^{\pm 1}$ is called *stabilizing* the braid and is referred to as the *stabilization move*.

Note that stabilization changes the number of strings in the braid by one and that we refer to both the addition and removal of a string as stabilizing. Starting with $e \in B_1$, we stabilize once to obtain $\sigma_1 \in B_2$. In this way, we can move between two braids in different braid groups. Markov's theorem states that conjugacy and stabilization are enough for knot equivalence.

Theorem 2.25 (Markov [55], Birman [14]). *Given two braids $\alpha \in B_n$ and $\beta \in B_m$, we have $\overline{\alpha} \approx \overline{\beta}$ if and only if β may be obtained from α by a finite sequence of conjugacy or stabilization moves; we denote this by $\alpha \approx_M \beta$ and call α and β Markov equivalent.*

In [55], Markov stated this theorem but the first complete published proof appeared in [14]. We shall not prove this theorem here as it is difficult and lengthy and would distract from the flow of the chapter. It is clear that we are interested in approaching knot classification from a braid theory point of view.

Definition 2.26 (Markov problem). Given two braid words $\alpha \in B_n$ and $\beta \in B_m$, the *Markov problem* or *algebraic link problem* asks whether $\alpha \approx_M \beta$.

By Markov's theorem, a solution to the Markov problem would provide a solution to the knot classification problem. As the Markov problem subsumes the conjugacy problem, we are presented with a hierarchy of three combinatorial problems in group theory which would have significant weight if a solution were found. Solutions exist to the word and conjugacy problems and we shall develop new solutions in later sections. A solution to the Markov problem only exists in the sense that there exists a knot classification scheme based on 3-manifold classification. This is an indirect and unusable solution as it is exponential in the amount of computing time required as a function of the crossing number of the knots concerned.

As many braid words represent the same braid, we are naturally lead to ask for a particularly simple representative. A very natural definition of "simple" in the case of a braid word is the least length possible; the shorter the word, the simpler it is. As length has an obvious minimum, we formulate the minimal word problem.

Definition 2.27 (minimal word problem). Given a braid word $\alpha \in B_n$, the *minimal word problem* asks one to find a braid word $\alpha_m \in B_n$ such that $\alpha_m \approx \alpha$ and $L(\alpha_m) \leq L(\alpha')$ for any braid word $\alpha' \approx \alpha$. The word α_m is the *minimal word* as it is, by definition, the shortest word in its equivalence class.

The word problem is frequently solved by devising an algorithm which finds a unique representative for the equivalence class of the element under consideration. In many groups, this so called *normal form* is also of minimum length in the equivalence class so that word problem and minimal word problem are solved together. Examples of such groups include free groups,

HNN-extensions and free products. For the braid groups, we shall find that the minimum word problem is far more complicated to solve than the word problem. In fact, under some very reasonable assumptions, the minimum word problem in the braid groups can only be solved by what amounts to a global search of the equivalence class. We shall outline these assumptions and the methods by which the global search may be done in later sections.

3 Braids and Knots

3.1 Notation for Knots

3.1.1 Dowker-Thistlethwaite Code

When he originally invented knot theory, Tait sought to construct a table of distinct prime knots. Two tasks had to be undertaken: (1) An exhaustive list of all possible knots had to be constructed and (2) all duplicates had to be struck from that list. Tait used mainly combinatorial methods to construct an exhaustive list and then tried to eliminate duplicates by trying to deform knots into each other. Tait labelled each crossing by a letter of the alphabet. He named a knot by starting at some random point and going along it in the direction of its orientation, writing down the label of the crossings as he passed them. This string of letters is called the *Tait code* of a knot. Clearly a knot with n crossings has a name consisting of $2n$ letters. There exist a finite, though large, number of possible such names for any n. Not all possible names can arise from naming a knot, however and Tait was able to find methods to determine if a specific name was valid.

Dowker and Thistlethwaite improved Tait's methods and introduced their own naming convention [35]. Consider a knot of n crossings and start at a random point going along the knot in the order of its orientation. Name the crossings in numerical order as you pass them giving each crossing two numerical labels (you will pass each crossing twice before returning to the staring point). Each crossing will have two numbers associated with it, i and j, say. We can construct an involution a from these by putting $a(i) = j$ and $a(j) = i$. Thus we have $a(a(k)) = k$ for any label k. Note that a reverses parity, that is if i is odd and $a(i) = j$, then j is even and vice-versa. We agree to write a_i for $a(i)$, S for the sequence $a_1, a_2, \cdots, a_{2n}$ and S_{odd} for the sequence $a_1, a_3, \cdots, a_{2n-1}$. The sequence S_{odd} completely determines both S and a [35].

There are $2n$ different possible starting points on the knot and if it has no orientation, there are two possible orientations. Not all sequences S_{odd} are identical. The *standard sequence* S_{std} for a knot will be the lexicographical minimum over all possible S_{odd}. Consider, for example, the trefoil knot in

Fig. 14 The procedure
for obtaining the Dowker-
Thistlethwaite code for the
trefoil.

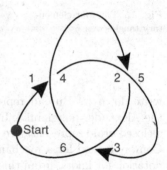

figure 14. The starting point is labelled in the figure and we proceed to name
the three crossings twice each in the order in which we encounter them. This
gives rise to the involution

$$a_1 = 4, \ a_2 = 5, \ a_3 = 6, \ a_4 = 1, \ a_5 = 2, \ a_6 = 3 \tag{35}$$

which results in $S_{std} = 4, 6, 2$.

Dowker and Thistlethwaite were able to determine both necessary and
sufficient conditions for a sequence of number to be realizable as a knot. This
algorithm is relatively simple and quick to implement and has been used
by them to tabulate knots. This code is very compact and easy to obtain
from a knot, but their tabulation methods focus on enumerating all possible
sequences and so we ask: How may we recover the knot given the code?

We note first of all that a knot of n crossings will get $2n$ labels ($2n$ being
necessarily even). Suppose $S_{std} = b_1, b_2, \cdots, b_n$ and the original involution
is a. As S_{std} is a standard sequence, we have that $a_{2i-1} = b_i$ for i ranging
from one to n. Because of the definition of a, we then have

$$a(a(2i-1)) = a(b_i) = 2i - 1 \tag{36}$$

Since a reverses parity and we list only the a_i for i odd in the standard
sequence, all these a_i are even. Thus we have recovered the entire involution a.
The crossings of the knot thus get the double names i and b_i for i ranging
from one to n. Having gotten the complete labelling information, we can
draw the crossings on our paper as double points and connect them in order
of the labelling and thus retrieve the knot. We may not get the same number
of crossings that we had in the original projection but this does not matter
topologically.

3.1.2 Conway's Basic Polyhedra

We constructed the Alexander polynomial in section 2.5. Starting from a
knot, we must first construct a closed braid isotopic to the knot, then

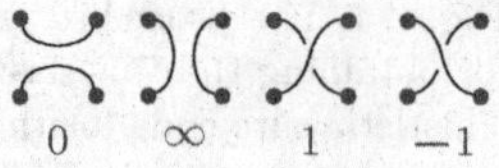

Fig. 15 The four elementary tangles.

write down its Bureau representation and take its determinant to obtain
the Alexander polynomial. In practise this process, particularly constructing
a closed braid representative, is very complicated and time-consuming. When
Conway looked for a mechanizable method, he was lead to construct a new
notation for knots. From this notation, he was able to extract the Alexander
polynomial so straightforwardly that he proceeded to calculate them all by
hand rather than mechanize the method.

A knot diagram has crossings and arcs connecting the crossings. If we
were to draw small circles around the crossings and then ignore what is in
the circles, we would have a template for the knot. Into the circles we could
insert any of the four elementary tangles of figure 15 to generate several
knots, one of which would be the original knot. The numerical names of
the elementary tangles arise from their classification which will not concern
us here. Conway's notation derived from the observation that many knot
diagrams are the same after the crossings have been so removed. In such
a way, we may generate a large number of different knots starting from one
such knot template and inserting different tangles into different slots. Conway
called these templates *basic polyhedra* and was able to show that eight basic
polyhedra are enough to generate all the different knots up to and including 11
crossings [30].

The number of different basic polyhedra needed to generate the knots of
higher crossing number n increases sharply with n and as the polyhedra lack
pattern, the next one may not easily be generated from the previous ones.
This problem gave rise to the idea of the universal polyhedron to be discussed
next.

3.1.3 The Universal Polyhedron

The *universal polyhedron* $P(i, j)$ is defined by figure 16. It has i rows and j
columns of *vertices* which will be filled with elementary tangles and which
are connected by *edges*. It can be shown that any knot can be represented by
some $P(i, j)$ and that we may write this down in a matrix form,

$$P(i, j) = \begin{pmatrix} p_{11} & p_{12} & \cdots & p_{1j} \\ p_{21} & p_{22} & \cdots & p_{2j} \\ \vdots & \vdots & \ddots & \vdots \\ p_{i1} & p_{i2} & \cdots & p_{ij} \end{pmatrix} \tag{37}$$

Fig. 16 The universal
polyhedron.

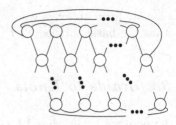

with the elements $p_{kl} \in \{1, -1, 0, \infty\}$. For example, the knots up to and
including six crossings (see figure 8) are given by

$$0_1 = (\infty) \qquad 3_1 = \begin{pmatrix} -1 & -1 \\ -1 & \infty \end{pmatrix}$$

$$4_1 = \begin{pmatrix} 1 & 1 & 1 \\ 0 & -1 & \infty \end{pmatrix} \qquad 5_1 = (1\ 1\ 1\ 1\ 1)$$

$$5_2 = \begin{pmatrix} \infty & 0 & 1 \\ -1 & -1 & -1 \\ 0 & 1 & \infty \end{pmatrix} \qquad 6_1 = \begin{pmatrix} \infty & 0 & -1 & -1 \\ 1 & 1 & 0 & \infty \\ 0 & -1 & -1 & \infty \end{pmatrix}$$

$$6_2 = \begin{pmatrix} 1 & 1 & 1 & \infty \\ 0 & -1 & -1 & \infty \\ 0 & 1 & \infty & 0 \end{pmatrix} \qquad 6_3 = \begin{pmatrix} 1 & 1 & 1 \\ 0 & -1 & \infty \\ 1 & 1 & 0 \end{pmatrix}$$

Theorem 3.1. *Every regular projection of any knot may be represented by
the universal polyhedron $P(i, j)$ for some i and j all the vertices of which
contain elementary tangles.*

Proof. A regular projection of a knot is characterized by a finite number n
of double points and $2n$ arcs which connect the double points in a specific
manner. For sufficiently large i and j, the polyhedron $P(i, j)$ can accommo-
date all double points in the form of ± 1 tangles and can achieve the desired
connection of these by placement of 0 and ∞ tangles into it. This is obvious
because the 0 and ∞ tangles represent horizontal and vertical connectors in
the polyhedron. Because this connection may be achieved without ± 1 tangles, it is clear that no further components, with the possible exception of
unknots, are created. Thus what remains to be shown is that no unwanted
unknots will be created.

   There are ij vertices and $2ij$ edges connecting them in the empty polyhedron $P(i, j)$. Eliminating one vertex by a 0 or ∞ tangle, eliminates two edges.
Apart from the ± 1 tangles of which there are n, the final polyhedron will contain $ij - n$ tangles of type 0 and ∞ which will have eliminated $2(ij - n)$ edges
from the original polyhedron, leaving exactly $2n$ edges which are needed to
connect the double points. Thus there is no extra edge left over which could

possibly form an extra component. Therefore any knot may be represented using the basic polyhedron $P(i,j)$ and elementary tangles. □

3.2 Braids to Knots

Suppose we have a braid b given by a braid word and we want to denote the knot that is isotopic to its closure in the standard notations. The universal polyhedron (figure 16) turned through π radians becomes a closed braid template if every vertex is filled with tangles of the types 1, −1 and 0.

The n-braid b will be specified by a function $b(t) = \sigma_i^{\pm 1}$ which gives the tth Artin generator of b for $1 \leq t \leq c$ where c is the number of crossings in the braid. The map ξ_i will map an Artin generator $\sigma_j^{\pm 1}$ to the elementary tangles 1 and −1 if the exponent of the Artin generator is 1 and −1 respectively and $i = j$ and will map any Artin generator to the elementary tangle 0 otherwise,

$$\xi_i(\sigma_i) = 1 \tag{38}$$

$$\xi_i(\sigma_i^{-1}) = -1 \tag{39}$$

$$\xi_i(\sigma_j^{\pm 1}) = 0 \text{ for } i \neq j \tag{40}$$

The closed braid $\bar{b}$ can be contained in the polyhedron $P(n-1,c)$ with $p_{ij} = \xi_i(b(j))$ with $1 \leq i < n$ and $1 \leq j \leq c$.

For example, $b = \sigma_1^3$. Then $b(i) = \sigma_1$ for $i = 1,2,3$ and we get

$$\overline{\sigma_1^3} = (1 \quad 1 \quad 1) \tag{41}$$

which correctly represents the trefoil knot. In the next section we will generalise this example to an infinite family of knots known as torus knots.

3.3 Example: The Torus Knots

The torus knots are an infinity family of prime knots which have particularly simple properties and are frequently used as examples in knot theory texts. The connection between braids and knots is readily illustrated in the case of torus knots and this is what we shall do below.

Definition 3.2. Given two co-prime (no common factors apart from unity) integers p and q, the torus knot $T_{p,q}$ is constructed by wrapping a closed curve around the surface of a torus such that it encircles it p times meridionally and q times longitudinally (respectively the short and the long way around).

Torus knots are completely characterized by the two integers p and q. They are invertible (isotopic under switching the orientation) and chiral (*not* isotopic to their mirror image). The fundamental group of their complements is given by $\pi_1(T_{p,q}) = \langle \{a, b\} : a^p = b^q \rangle$ [49] from which we may recognize the requirement that p and q be co-prime. It may be shown that $T_{p,q} = T_{q,p}$. The torus knots are among the few knots for which the minimal number of crossing-switches required to transform the knot into the unknot, i.e. the *unknotting number*, is known; it is $(p-1)(q-1)/2$ [1]. They are also among the few knots for which the minimal number of crossings in any projection, i.e. the *minimal crossing number* is known; it is $\min[p(q-1), q(p-1)]$ [75].

The simplest example of a torus knot is the trefoil. It is not the unknot even though $T_{p,1}$ for any p would be isotopic to the unknot but p and unity are not coprime. The trefoil knot is $T_{3,2}$ and the knot 5_1 is $T_{5,2}$. In general, the closed p-braid $(\sigma_{p-1}\sigma_{p-2}\cdots\sigma_1)^q$ is isotopic to the torus knot $T_{p,q}$. This is easily seen by picturing $T_{p,q}$ on an actual torus. We cut the torus across a random meridian (the short way around) and straighten it into a cylinder. The remainder will be the braid above. Moreover, there exists no closed braid representative of $T_{p,q}$ with less than p strings so that p is the *braid index* of $T_{p,q}$.

Exercise 3.3. Show that the Alexander polynomial of $T_{p,q}$ is given by

$$\triangle_{T_{p,q}} = \frac{(1-t^{pq})(1-t)}{(1-t^p)(1-t^q)} \tag{42}$$

As we have seen, any torus knot can be represented as the p-braid $b = (\sigma_{p-1}\sigma_{p-2}\cdots\sigma_1)^q$. By using the method of the last section, we can represent $T_{p,q}$ in the polyhedron $P(p-1, q(p-1))$.

$$T_{p,q} = \begin{pmatrix} & 1 & & 1\cdots & & 1 \\ & \cdots & & \cdots & \cdots & & \cdots \\ & 1 & & 1 & \cdots & 1 \\ 1 & & 1 & & \cdots & 1 \\ 1 & & 1 & & \cdots 1 \end{pmatrix} \tag{43}$$

The Dowker code may be obtained from a knot represented in our notation by simply walking through the polyhedron and labelling the crossings. As the polyhedron is structured, this walk is perfectly definite and can be programmed easily on a computer.

3.4 Knots to Braids I: The Vogel Method

A braid is more structured than a knot and so the transition from knot to closed braid is harder to effect than the reverse. There exists a simple

Fig. 17 Both types of
crossing have to be recon-
nected in the shown way in
order to obtain a diagram
of Seifert circles from a
knot diagram.

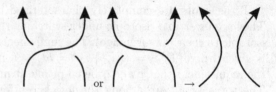

method due to Vogel [72] which we shall present without proof in this section
Suppose we are faced with a knot diagram D which we want to convert
into a closed braid. For this method, we shall have to view D in a variety
of ways. From the diagram, we can get to the projection P of the knot
onto the plane by viewing each crossing as a double point and thus ignoring
over and undercrossing information. From D, we can construct a diagram
S by reconnecting each crossing in D in the manner shown in diagram 17.
The diagram S will contain a number of unknots which we will call *Seifert
circles*. Using these constructions, we can define the crucial concept in Vogel's
method.

Definition 3.4 (admissible triple). Let f be a face of P and a and b be
two edges of P. The triple (f, a, b) is called an *admissible triple* if and only
if it satisfies: (i) a and b are contained in different Seifert circles and (ii) a
and b have the same orientation with respect to any orientation of ∂f, the
boundary of f.

It is shown in [72] that the following algorithm will transform any knot
diagram D into a diagram of a closed braid.

Algorithm 3.5 *Input: A knot diagram D. Output: A knot diagram D'
ambient isotopic to D and in the form of a closed braid.*

1. Determine if D has an admissible triple. If yes, continue. If no, D is in the
   form of a closed braid and the algorithm is done.
2. Admissible triples can come in the two flavours shown in the solid curves
   in figure 18. Each admissible triple detected, is to be transformed (via a
   Reidemeister move type 1) from the solid curve to the dashed curve in figure 18. Such a transformation will be called an *elementary transformation*.
   Then go back to step 1 of the algorithm.

It can be shown that algorithm 3.5 always terminates after at most
$(s - 1)(s - 2)/2$ elementary transformations where s is the number of
Seifert circles in S. The braid which this algorithm generates has at most
$n + (s - 1)(s - 2)$ crossings where n is the number of crossings in D as the
elementary transformation adds two crossings each time it is applied. It is
unclear whether algorithm 3.5 is confluent, that is whether the order in which

Fig. 18 The two types of
admissible triples are shown
in the solid curves and the
form into which they should
be transformed is shown in
the dashed curves.

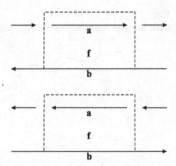

we perform elementary transformations should two (possibly overlapping) triples be simultaneously admissible changes the final outcome.

3.5 Knots to Braids II: An Axis for the Universal Polyhedron

Having constructed a new notation for knots, we wish to solve the problem of how to extract a closed braid from the matrix which is isotopic to the knot described by the matrix. A few algorithms have been constructed in the past, which convert a knot into a closed braid but they are difficult to implement because they depend upon topological deformation of the knot projection [51] [15]. The best known algorithms have been implemented [72] [77] and have complexity $O(n^2)$. We shall present an algorithm which achieves the conversion with complexity $O(n)$, increases the number of crossings only in a few cases (and then only by a few crossings) and uses a linearly bounded number of strings. There exists no algorithm to calculate the number of strings which are at least necessary to describe a specific knot — the braid index of the knot. Because of this, it is not possible to say how close to the minimum the number of strings used by our algorithm is. The number of crossings is sometimes increased because it has been found that there are knots for which any closed braid representative has more crossings than the minimal knot diagram; the knot 5.1 in the standard tables is the simplest example of this [66]. Our algorithm is valid both for oriented and unoriented knots.

3.5.1 An Example

Alexander's theorem was proven by showing that every knot can be deformed into a form where the knot loops around an axis a finite number of times without local maxima or minima with respect to that axis. If we cut the string along the axis in one place, we obtain a braid. The gluing back of the

Fig. 19 The trefoil knot
with an axis for braiding it.

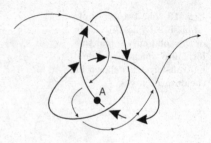

Fig. 20 The trefoil knot
as it appears after the axis
has been straightened from
figure 19. For reference the
point A has been labeled
here again.

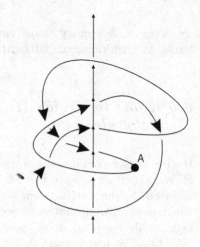

cut constitutes the canonical closure. Thus as far as the canonical closure
is concerned, the finding of an appropriate axis is the key. Having obtained
a canonically closed braid which is equivalent to a knot, we may obtain a
plait from it by considering the closure curves part of the braid diagram and
moving them into the middle of the braid diagram. The next section gives
an example of this.

For the rest of this section, we are going to work through an example
of our method. Consider the trefoil knot in figure 19. We have drawn an
axis through it by the following method: (1) We drew a line through the
projection of the trefoil which intersects every region of the plane at least
once, (2) begins and ends in the infinite region and then (3) assigned the
under and overpasses of the knot under and over the axis by traversing the
knot from a random starting point (point A in the figure) while (4) assigning
the passes alternately as we met the crossings of axis and knot. Next we
perform a coordinate transformation from the knot reference frame (figure
19) to the axis reference frame in figure 20 by pulling the axis straight.

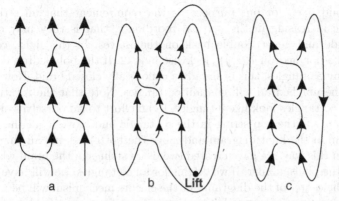

Fig. 21 The braid which is extracted from figure 20 by cutting the trefoil knot at its overcrossings over the axis and laying out the ends is displayed in part (a). The closure of this braid is part (b). If we lift the arc labeled in part (b) we obtain the braid in part (c). See discussion in the text.

We can easily observe from figure 20 that the axis is valid; i.e. if we traverse the knot starting at A we will travel around the axis without local maxima or minima permanently in a clockwise direction. If we now cut the knot at those points at which it overcrosses the axis and lay out the ends carefully to either side, we shall obtain the braid $\sigma_1^{-1}\sigma_2^{-1}\sigma_1^{-1}\sigma_2^{-1}$ shown in figure 21 (a). To get back to the trefoil from this, we perform the canonical closure which is identical to sealing the cuts made above. This is shown in figure 21 (b). This knot has four crossings and is ambient isotopic to the trefoil thus there is some inefficiency in our braid representation (note however that there exist knots for which the most efficient braid representation contains more crossings than their most efficient knot projection [66]). We note that we may lift the arc labeled in figure 21 (b) to remove one crossing. This move also removes a string and so we obtain the braid of figure 21 (c). This braid has two strings and three crossings, it is thus the most efficient representation of the trefoil as the trefoil must have at least this many strings and crossings. We conclude that the closure of the braid $\sigma_1^{-1}\sigma_1^{-1}\sigma_1^{-1}$ is ambient isotopic to the trefoil knot. Note that we may turn the entire figure 21 (c) about a vertical axis through its center and thus obtain the result that the braid $\sigma_1\sigma_1\sigma_1$ is ambient isotopic to the trefoil also; this, finally, is the well-known braid representation of the trefoil knot. This is the prototype for a general method which we shall develop below.

3.5.2 Platting a Knot

The diagram of a knot which is expressed as a closed braid may be naturally divided into two parts: the braid and the closure. The most important feature of the braid part, for our purpose, is the requirement that all strings be monotonic increasing in the vertical coordinate, that is they may only go side to side and never double back on themselves. In this light, consider turning the polyhedron $P(i,j)$ clockwise by $\pi/2$. If the polyhedron does not contain any ∞ tangles, this is already a canonically closed braid. However, in general, the polyhedron will contain ∞ tangles. Note that the rotation will make the ∞ tangles look like 0 tangles. In an effort to rid ourselves of the ∞ tangles, we take the top string in the ∞ tangle and move it all the way to the bottom of the knot diagram and move the bottom string all the way to the top. In this way, we have created two extra strings in the braid which are closed in the plait manner. If we do this for all ∞ tangles, we will have a valid braid in the center of the diagram but the closure mechanism will be a hybrid between the canonical and plait methods. In order to rectify the situation, we move the strings which are closed in a canonical manner into the center of the braid diagram, thereby creating more strings and more crossings. Once this has been done, we have a fully valid braid closed in the plait manner which is ambient isotopic to the knot we started with. Figure 22 shows the process of converting the unknot

$$U = \begin{pmatrix} -1 & 1 \\ \infty & -1 \end{pmatrix} \tag{44}$$

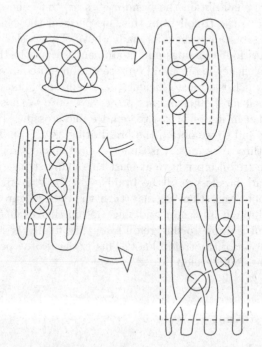

Fig. 22 The conversion of a knot into a plait.

into the braid $\sigma_2\sigma_4^{-1}\sigma_3\sigma_4\sigma_5^{-1}\sigma_6^{-1}\sigma_4^{-1}\sigma_5^{-1}\sigma_4\sigma_6$ closed in the plait manner. This procedure is valid generally and clearly represents a readily implementable algorithm for transforming a knot given in our notation into a plait. If the original knot is given in the polyhedron $P(i, j)$ and has k tangles of type ∞, then the number of strings required in the plait is $2(i + k + 1)$ but the number of crossings depends upon the exact configuration.

3.5.3 Laying the Axis

As mentioned before, the transformation of a knot projection into a canonically closed braid centers around finding an appropriate axis for the string to wind around. This was the central point of Alexander's theorem which proves that such an axis may always be found. A ready method for finding an axis is given in the following algorithm.

Algorithm 3.6 *Input: A knot projection. Output: A knot projection with an axis around which the knot winds without local maxima or minima.*

1. Begin with enumerating the regions into which the knot projection divides the plane, suppose there are R of these.
2. Choose two arbitrary points in the infinite region and call them A and B.
3. Draw a line L connecting A and B in such a way that the line intersects every region at least once.
4. Choose a random point on each of the knot's components and traverse the knot in the direction of the orientation once for each component starting at the chosen point. While traversing label each intersection of L with the knot alternatingly with a $+$ or $-$ sign starting with $+$.
5. Interpret each $+$ crossing as an overcrossing of L over the knot and each $-$ crossing as an undercrossing of L under the knot. The line L oriented from A to B is then a valid axis.

This algorithm may clearly be applied to our polyhedron $P(i, j)$. However we have the problem of the regions which depends upon the exact configuration of the knot. This can be solved by forcing the line L to intersect every region in the *polyhedron* and therefore intersecting some regions of the *knot* more than once. This is unfortunate but unavoidable if we are seeking a general solution of the problem. The manner in which this may be done most economically is illustrated in figure 23. The line L is the dotted line beginning at point A and finishing at point B. If the polyhedron has an odd number of columns (as the one in figure 23), then the line L is best described by the dotted line in figure 23. If however, the polyhedron has an even number of columns, then the line L is best described by the dotted line in figure 23 from point A to point C and then the dashed line from point C to point B.

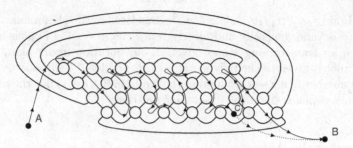

Fig. 23 The axis of the braid through the polyhedron $P(i, j)$.

If algorithm 3.6 is correct then a line drawn in a general polyhedron $P(i, j)$ according to this example is a valid braiding axis.

We may find an axis which passes through every region exactly once, if possible, by the following algorithm.

Algorithm 3.7 *Input: A matrix describing a knot in our notation. Output: A matrix describing the regions of the knot. Each element of the matrix receives a label from 1 to R, the number of regions. This gives complete information about which regions of the polyhedron are connected and how many there are.*

1. Begin at the top left of vertex $(1,1)$ and follow the boundary downwards, as for counting regions, the orientation of the knot does not matter. Mark the region $(0,1)$ with a 1, the current marker, in the region matrix.
2. In following the boundary, one will come to vertex $(1,1)$; we assess its value and continue. If we stay in the same region of the *polyhedron* we continue, if we enter a new region of the polyhedron, then this new region of the polyhedron belongs to the same region of the *knot* as the previous one and thus we mark it with the current marker in the region matrix. The whole issue at hand is that the regions of the polyhedron are known while we wish to gain knowledge of the regions of the knot.
3. We continue to follow the boundary until we reach the point of origin.
4. We search the matrix for an unmarked region. If there exist unmarked regions, we increment our current marker and choose one of the regions as our new starting region and choose a point upon its boundary as our new starting point. Then, we repeat the algorithm from step 1, marking the region with the current marker.
5. Once no unmarked region of the polyhedron exists, the algorithm is finished. The largest marker used in the matrix which we have obtained is clearly the number of regions of the knot. Furthermore, since all connected regions are labeled with the same marker, we have a complete knowledge of which regions of the polyhedron belong to the same region of the knot.

Algorithm 3.8 *Input: A knot projection given in our notation. Output: An axis which passes through every region exactly once, if this is possible. If not the output is an axis which passes through each region at least once.*

1. Get the region information as prescribed in algorithm 3.7.
2. Construct a graph in which each region is symbolized by a node and two nodes are connected by an unweighed edge if they are adjacent in the plane.
3. A Hamiltonian circuit is then a path which passes through each region, that is node, exactly once starting in the infinite region and returning there. If a Hamiltonian circuit exists, so does an optimal axis. If no Hamiltonian circuit exists, we find an axis using algorithm 3.6 which gives an axis which passes through every region at least once.

The advantage is that we will generate a braid with less strings but the Hamiltonian circuit problem is NP-complete and so the execution of algorithm 3.8 is exponential (unless we use an approximation algorithm or it is shown that P = NP). This fact lends further weight towards the usefulness of algorithm 3.6. The primary usefulness of this algorithm originates in the fact that the laying of the axis does not depend upon the exact knot configuration, only the labelling does. Before we continue, we prove that algorithm 3.6 always yields a valid axis, this essentially amounts to proving Alexander's theorem.

Theorem 3.9. *Given any knot projection, algorithm 3.6 will find an axis about which the knot is without local maxima or minima.*

Proof. Alexander's theorem [4] states given a knot projection, it is possible to deform it with respect to a point P in the projection plane that after the deformation a point A which travels along the knot in the direction of its orientation will travel around the axis defined by P (the axis is a line perpendicular to the projection plane intersecting it at P) in a constant fashion, either clockwise or counterclockwise, for the entire circumnavigation of the knot. We wish to do the opposite, namely to deform the axis around the knot projection to achieve the same ends. We can imagine the process of laying the axis as akin to sewing in which we move the needle up from and down onto the plane. Morton [59] has constructed a similar method to ours which he calls "threading."

The knot divides the plane into several regions. If the axis does not intersect a particular region, the point A will change course during traversing the knot and so the axis must intersect each region. It is however clearly only necessary for the axis to intersect the region once. Choose a line in the plane which intersects the axis. With respect to this line we can define an angular coordinate θ going around the axis. As point A must travel around the axis in a constant fashion it must, after it passes $\theta = 0$, reach $\theta = \pi$ before it once again reaches $\theta = 0$. This shows that the axis, in the projection plane, must over and undercross the knot alternately with respect to A. This fulfills the requirements of an axis and these are assured by algorithm 3.6 and thus the theorem is proven.   □

3.5.4 Getting the Braid

Having obtained the axis, we must now simply put together all the pieces and construct the braid. This will be done via the following algorithm.

Algorithm 3.10 *Input: An axis L in a knot projection given in $P(i,j)$ using our notation. Output: A braid the canonical closure of which is ambient isotopic to the given knot.*

1. Consider an empty polyhedron $P(i,j)$ and label each edge by the row and column index of the vertex out of which it is emerging on the right side giving it the further label a if it is the top edge and b if it is the bottom edge. That is the top right hand edge coming out of the vertex $(1,1)$ would be $(1,1)_a$.
2. All edges which intersect the axis L at a positive crossing are to be numbered in order starting at point A; suppose there are k of these.
3. Starting at the numbered edges, use the traversal algorithm to follow each edge around the knot until another positive crossing with the axis L. All edges encountered are to be labelled with the same number as the original edge.
4. When all edges are numbered, we have identified the individual strings of the braid and numbered them in order. Assign a distance value of 1 to each edge in the polyhedron.
5. Traverse the knot again as in step 3 but this time stopping at each double point and extracting which labelled string passes over which other labelled string and at which distance value this occurs.
6. When the whole has been traversed, we have a list of crossings specifying which strings are involved, which string crosses over the other and at what distance from the bottom of the braid the crossing occurs. This information may be used readily to construct a colored braid, which may be converted easily into an Artin braid word.
7. We assess the string labels around the knot and calculate the permutation associated with the braid which winds around our axis. If this permutation is different from the permutation of the braid which we obtained in step 6, the residual permutation must be added to this braid in the form of extra crossings.

The number of crossings is increased in some circumstances by a small amount in step 7 of the algorithm. It is a fact that there exist knots of minimal crossing number n which have closed braid representatives all of which have crossing numbers greater than n [66]. Hence, step 7 is not a deficiency of the algorithm 3.10 but a fundamental necessity.

It is clear from Alexander's theorem[4] that this algorithm works. The number of strings used is the number of positive crossings of the axis with the knot which is equal to half the number of crossings. The number of crossings of the axis with the knot is

$$N_c = \begin{cases} 4i + (2i+2) \left\lfloor \frac{i-2}{2} \right\rfloor & j \text{ odd} \\ 2i + (2i+2) \left(\frac{i-2}{2} \right) + 2 & j \text{ even} \end{cases} \tag{45}$$

where $\lfloor x \rfloor$ is the greatest integer less than x. An analysis of the possibilities in oddness and evenness of i and j reveals that N_c is always even which is good since we must have an equal number of positive and negative crossings.

Algorithm 3.10 therefore finds a braid with a number of strings which scales linearly in the number of rows and columns necessary to represent the knot. It is conceivable that a more economical way of laying an axis may be found using algorithm 3.8 but this has an exponential complexity. The number of strings may be reduced after the braid has been found using Markov's theorem.

The determination of the regions, the laying of the axis, the labelling of the axis crossings, the labelling of the edges and the extraction of the double point information all take a time proportional to the number of vertices in the polyhedron ij. The building of the braid from the crossing information takes time proportional to ij. Therefore the entire algorithm to proceed from a knot projection to a canonically closed braid has complexity $O(ij)$. This algorithm succeeds in being readily implementable and in constructing a braid which is reasonably small.

3.6 Peripheral Group Systems of Closed Braids

In this section we will investigate the peripheral group system of the closure of the fundamental braids. The peripheral group system is a complete invariant of knots and figures largely in knot theory. The fundamental braid words are very important in braid theory and we choose them for this investigation for that reason and that the closure of Δ_3 is the Hopf Link and the closures of the other fundamental braids look very similar to Hopf Links. In fact so similar that we can regard the class of knots defined by the closure of fundamental braids as a generalization of the Hopf Link. Another generalization of the Hopf link has been investigated in the literature [28]. We shall see that our methods developed here can be extended beyond fundamental braid words to all braid words.

3.6.1 The Fundamental Group

Consider a space X and a point $x_0 \in X$ which we shall call the *base point*. In the space X, we may construct loops, i.e. paths from x_0 to itself.

Definition 3.11 (fundamental group). The group that consists of the loops at x_0 in the space X with respect to the homotopy equivalence relation (continuous maps from one set of loops to another) is called the *fundamental group* of the space X and is denoted by $\pi_1(X, x_0)$.

It can be shown that if X is path connected, the choice of base point does not matter. All spaces we are about to consider are path connected and so we shall drop the specification of the base point and denote a fundamental group by $\pi_1(X)$.

Definition 3.12 (knot complement). The *complement* of a knot K with respect to a space X is the space $X - V(K)$ where $V(K)$ represents a tubular neighborhood of the knot K.

Definition 3.13 (knot group). The *group* of a knot K, denoted by $\pi(K)$ is the fundamental group of the complement of the knot with respect to the space $X = \mathbb{R}^3$ (sometimes X is taken as S^3 but it can be shown that the two fundamental groups arising from $X = \mathbb{R}^3$ and $X = S^3$ are isomorphic).

Consider a knot K and its complement $\mathbb{R}^3 - V(K)$. In this space, choose a point x_0 as the base point (the choice is arbitrary as the space is path connected). Consider the projecting cylinder $Z \in \mathbb{R}^3$ which contains all of $V(K)$ and is constructed such that the projection of K onto the plane $z = 0$ in $\mathbb{R}^3$ contains at most double points. This projecting cylinder will contain n self-intersections if K has n crossings. Label these self-intersections by a_i and the sections of Z between the a_i by s_i. Z is to be oriented in order to match the orientation of K. Denote a loop from the base point to itself which goes exactly once around s_i and no other s_j by p_i. From this construction, it is clear that all loops (with respect to homotopy equivalence) can be constructed as products of the loops p_i. This can easily be seen by the fact that any path in the complement can be continuously deformed into a product of loops p_i. Thus $\pi(K)$ is generated by the loops p_i. For an example see figure 24.

Having gotten the generators, we need the relations to obtain a group presentation. Consider the loops encircling the a_i and join them to the base point by a path c_i, then it is clear that $c_i a_i c_i^{-1}$ is contractible (homotopic to the trivial loop). The word in the group corresponding to this loop is then a relation in the group. The presentation so obtained is called the *Wirtinger*

Fig. 24 The labelling of
the trefoil knot in order to
yield a presentation for the
fundamental group of its
complement.

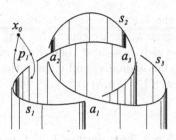

Fig. 25 The two possible
forms of double points in
the diagram of an oriented
knot. The crossing of type 1
has characteristic $\epsilon = 1$ and
the type 2 has characteristic
$\epsilon = -1$.

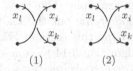

(1) (2)

presentation of the knot group. This discussion proves that the following
algorithm to obtain it is correct.

Algorithm 3.14 (Wirtinger presentation of the knot group)
Input: A knot projection of a knot K. Output: The Wirtinger presentation of
$\pi(K)$.

1. *Label all overcrossing arcs in the projection by g_i for $1 \leq i \leq n$ where n
   is the number of double points in the projection starting at any point and
   then assigning the labels in order corresponding to the orientation of the
   knot.*
2. *For each double point, determine its characteristic sign, see figure 25. The
   characteristic can be most easily determined by the "right hand rule" which
   says that if you spread your thumb at right angles from your fingers and
   point it along the overcrossing arc, and if your finger point along the undercrossing arc, the characteristic is 1 and -1 otherwise (along being taken to
   mean along the orientation of the knot).*
3. *$\pi(K) = \langle \{g_1, g_2, \cdots, g_n\} \mid \{r_1, r_2, \cdots, r_n\} \rangle$ where the relators are given by
   $r_j = g_j g_i^{-\epsilon_j} g_k^{-1} g_i^{\epsilon_j}$ and ϵ_j is the characteristic of the crossing associated
   with r_j.*

It can be seen that one relation can be derived from the others and
thus a knot group has deficiency one. The knot group is an invariant of
knots as is trivially seen by definition but it is not complete. Thus if two
knots have isomorphic knot groups they are not necessarily isotopic. However if they have non-isomorphic groups, the knots are distinct. The unknot
U has the knot group constructed by a single generator and no relations,
i.e. $\pi(U) = \mathbb{Z}$, the infinite cyclic group. It is a practical observation that

the Wirtinger presentation can often be simplified considerably in that some generators are removable [38]. In particular, $\pi(K)$ for the torus knot $T_{p,q}$, which is a knot that winds around a torus p times the short way around (meridionally) and q times the long way around (longitudinally), is given by $\pi(K) = \langle \{a, b\} : a^p = b^q \rangle$ [49].

3.6.2 The Square and Granny Knots

The knot group is not a complete invariant. We wish to illustrate this by an example. The knot groups of the square S and granny G knots are isomorphic but the knots are distinct. The demonstration of this fact will occupy this section. For a picture of these knots, see figure 26. We follow algorithm 3.14 to compute the knot groups. Both knots receive six generators and six relations. All crossings in the granny knot have characteristic -1 as well as three crossings in the square knot, the other crossings in the square knot have characteristic 1. We thus write down the Wirtinger presentations.

$$\pi(G) = \left\langle \begin{array}{l} \{g_1, g_2, g_3, g_4, g_5, g_6\} \mid g_6 g_2 = g_2 g_1;\ g_5 g_1 = g_1 g_6; \\ g_1 g_6 = g_6 g_2;\ g_2 g_4 = g_4 g_3;\ g_4 g_3 = g_3 g_5;\ g_3 g_5 = g_5 g_4 \end{array} \right\rangle \quad (46)$$

$$\pi(S) = \left\langle \begin{array}{l} \{g_1, g_2, g_3, g_4, g_5, g_6\} \mid g_6 g_2 = g_2 g_1;\ g_5 g_1 = g_1 g_6; \\ g_1 g_6 = g_6 g_2;\ g_3 g_2 = g_2 g_4;\ g_2 g_4 = g_4 g_3;\ g_4 g_3 = g_3 g_5 \end{array} \right\rangle \quad (47)$$

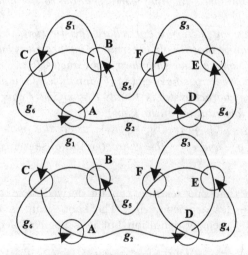

Fig. 26 The construction of the Wirtinger presentation of the fundamental group of the complement of the (a) square and (b) granny knots.

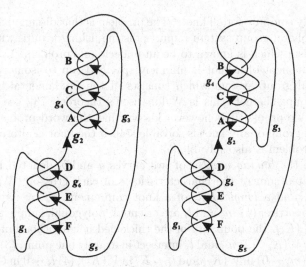

Fig. 27 The (a) square and (b) granny knots from figure 26 transformed into closed braids.

Three generators may be defined in terms of the other three in both groups and after some simple manipulation we obtain

$$\pi(G) \approx \pi(S) = \left\langle \begin{array}{c} \{g_1, g_2, g_3\} \mid \\ g_1 g_2 g_1 = g_2 g_1 g_2; \ g_2 g_3 g_2 = g_3 g_2 g_3 \end{array} \right\rangle \tag{48}$$

Even though the knot groups are isomorphic, the knots are distinct. This can seen by deforming the knots into closed braids and computing their Jones polynomials. The transformation is straightforward and the result is shown in figure 27. Reading off from the figure, we obtain that

$$G \approx \overline{\sigma_1^{-3}\sigma_2^3}; \qquad S \approx \overline{\sigma_1^3\sigma_2^3} \tag{49}$$

Recalling the algorithm to compute the Jones polynomial yields

$$V_G t = 3 - \left(t^3 + t^{-3}\right) + \left(t^2 + t^{-2}\right) - \left(t + t^{-1}\right); \qquad V_S(t) = \left(t + t^3 - t^4\right)^2 \tag{50}$$

and we see that the knots are distinct proving that the fundamental group is a incomplete invariant.

3.6.3 Peripheral Group System

The fundamental group is not a complete invariant but it is possible to refine our methods to construct a complete invariant from the fundamental group, the peripheral group system. This is the only complete invariant of knots that we can readily compute for all knots. The problem is that distinguishing knots via the peripheral group system requires distinguishing groups with respect to isomorphism which is known to be an undecidable problem [2, 3, 64].

The complement of a knot is uniquely specified (up to isomorphism) by its peripheral group system which consists of the fundamental group and a few subgroups thereof (this is Waldhausen's theorem [73], see [40] for a more accessible proof). It is however known that the word problem for any fundamental group of any knot is solvable [74]. If the knot is alternating, the conjugacy problem is also solvable [5].

We define the *linking number* of two curves a and b, denoted by $lk(a, b)$ as the weighted sum of the characteristics ϵ of each crossing. We define a *meridian* m_i and a *longitude* l_i of a knot component K_i by requiring the following properties: (1) m_i and l_i are oriented, polygonal, simple and closed curves in $\partial V(K_i)$, the boundary of the thickened neighborhood of K_i which we denote by $V(K_i)$, (2) m_i and l_i intersect in exactly one point, (3) m_i is null homologous ($m_i \sim 0$) in $V(K_i)$ and $l_i \sim K_i$ in $V(K_i)$, (4) $l_i \sim 0$ in $C(K_i)$ and (5) $lk(m_i, K_i) = 1$ and $lk(l_i, K_i) = 0$ in S^3. The above five properties define m_i and l_i uniquely up to isotopy on the boundary of $V(K)$ (see figure 28 for an illustration). The *meridian-longitude system pair* $\mathcal{M}(K)$ for a j-component knot K is the pair of sets $(\{m_1, m_2, \cdots, m_j\}, \{l_1, l_2, \cdots, l_j\})$. Interestingly, if the longitude is trivial (equivalent to the identity element) in the knot group, the group is infinite cyclic; i.e. the group is isomorphic to the fundamental group of the complement of the unknot. Furthermore, the meridians and longitudes commute with each other in any knot group [49].

The meridians can be taken as the Wirtinger generators of the knot group. The longitude of a knot may be read off the projection very easily. Begin with the first generator and traverse the knot in the direction of its orientation. Whenever undercrossing an arc, write down the generator of the undercrossed arc to the power of the negative of the characteristic of the crossing. After the

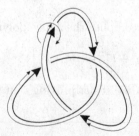

Fig. 28 The thick curve displays the trefoil knot with an orientation. The thin curve which is parallel to the trefoil knot is the longitude; the orientation of the longitude is the same as the knot. The thin curve encircling both trefoil and longitude at the top left hand corner is the meridian. Note that the five conditions given in the text are fulfilled by these curves.

full traversal, append as many copies of the initial generator (or its inverse) to make the total sum of exponents equal to zero.

The meridians and longitudes of $\mathcal{M}(K)$ may be considered to be elements of $\pi(K)$ by choosing a path p_i in $C(K)$ from the base point x_0 to the (unique by definition) point $m_i \cap l_i$ for each i. Then the subgroup $\langle m_i, l_i \rangle$ of $\pi(K)$, generated by m_i and l_i is independent of the choice of p_i up to conjugation. The *peripheral group system* of a j-component knot K is $p(K) = (\pi(K); \mathcal{M}(K))$. By an isomorphism ϕ between two peripheral group systems $p(K) \approx_\phi p(K')$, we mean $\pi(K) \approx \pi(K')$ such that $\phi(m_i) = m'_i$ and $\phi(l_i) = l'_i$ for all i. It can be shown that for any two knots K_1 and K_2, $p(K_1) \approx p(K_2)$ if and only if $K_1 \approx K_2$ [49]. If we restrict attention to prime knots of a single component, we have $\pi(K_1) \approx \pi(K_2)$ if and only if $K_1 \approx K_2$ [49]. Thus the problem of knot isotopy can be transformed into the problem of peripheral group system isomorphism. Since it is not possible to determine, in general, if two groups are isomorphic, this does not solve the knot classification problem.

4 Classification of Braids and Knots

In this section we will discuss the word, conjugacy, Markov and minimal word problems introduced in section 2.8 in detail. There are a number of distinct solutions to the word and conjugacy problems in B_n each of which has special features and gives additional insight into the problem and braids in general. We shall discuss briefly the solutions due to Garside as they are historically the most important ones and have had ground-breaking influence on the field. Then we shall discuss a new solution to both problems. The Markov problem is unsolved at present but we will discuss some of its features and why it is so difficult. Lastly, we will present a number of results about the minimal word problem. It is computationally expensive to solve (NP-Complete) and so we present a heuristic algorithm and some simulation approaches and interesting results arising from them.

4.1 The Word Problem I: Garside's Solution

In the Artin presentation, the word problem asks whether two n-braid words $\alpha = \sigma_{a_1}^{\epsilon_1} \sigma_{a_2}^{\epsilon_2} \cdots \sigma_{a_o}^{\epsilon_o}$ and $\beta = \sigma_{b_1}^{\eta_1} \sigma_{b_2}^{\eta_2} \cdots \sigma_{b_p}^{\eta_p}$ such that $1 \leq a_i, b_j < n$ and $\epsilon_i, \eta_j = \pm 1$ for $1 \leq i \leq o$ and $1 \leq j \leq p$ are equivalent to each other (denoted $\alpha \approx \beta$) under the equations $\sigma_i \sigma_j \approx \sigma_j \sigma_i$ for $|i - j| > 1$ and $\sigma_i \sigma_{i+1} \sigma_i \approx \sigma_{i+1} \sigma_i \sigma i + 1$. This problem is traditionally approached by trying to find an algorithm which will use the two braid group relations to construct, from any braid, a unique normal form. A *unique normal form* $\underline{\alpha}$ of the braid α and the corresponding form $\underline{\beta}$ of the braid β are exactly equal ($\underline{\alpha} = \underline{\beta}$) if and only if $\alpha \approx \beta$.

Artin was the first to describe a normal form to solve the word problem for braids [7] but the normal form has a exponentially growing number of crossings in terms of the original number of crossings and so is not as useful as other methods. Garside constructed another unique normal form which has many important properties [37]. Garside's method begins with the crucial proposition,

Proposition 4.1. *For any* σ_i^{-1}, *we have* $\sigma_i^{-1} \approx \Delta_n^{-1} \Delta_{n-1} d_{n-1,i+1} d_{i-1,1}$.

Together with the fact that $\sigma_i \Delta_n \approx \Delta_n \sigma_{n-i}$ (see proposition 2.19), this means that we may put any n-braid word α into the form $\alpha \approx \Delta_n^{-q} \alpha'$ where α' is a positive braid word and q the number of inverse generators present in α. Next we need to construct the diagram $D(\alpha')$ of the positive braid α'.

Definition 4.2. The *Cayley diagram* $D(b)$ (or *diagram* for short) of a positive braid word b is the set of all braid words equivalent to b.

Because of proposition 2.21 which states that two positive and equal braid words are positively equal, the following algorithm obviously constructs $D(b)$ given a positive b.

Algorithm 4.3 *Input: A positive braid word b. Output: The diagram $D(b)$ of b.*

1. We define the set $D_0(b)$ by $D_0(b) = \{b\}$.
2. The set $D_i(b)$ is obtained from the set $D_{i-1}(b)$ by adding to $D_i(b)$ all the braid words which can be obtained from any member of $D_{i-1}(b)$ by applying any braid group relation exactly once and the deleting all those which are already contained in the sets $D_j(b)$ for $0 \le j < j$.
3. Step 2 is recursively applied until an integer m is reached for which $D_m(b) = \emptyset$. It is obvious from the construction of the sets that a finite m always exists.
4. The set $D(b)$ is then the union of the D_i for $0 \le i \le m$.

We note that, in general, $D(b)$ contains a number of elements which grows exponentially both with the length and number of strings in b. Once $D(\alpha)$ has been fully enumerated we select from it a word of the form $\Delta_n^p \alpha''$ for which p is maximal. Then we construct $D(\alpha'')$ and choose the braid word α''' from $D(\alpha'')$ which has the lowest integer associated with it, the decimal expansion of which is given by the concatenation of the generator indices in the braid word ($\sigma_2 \sigma_1 \sigma_3$ has associated integer 213). The braid word α''' is called the *base* of the diagram $D(\alpha'')$. We define the *Garside normal form* of the braid α to be $\alpha_G = \Delta_n^{p-q} \alpha'''$. We refer to $p - q$ as the *Garside exponent* and to α''' as the *Garside remainder*. It can be shown that the Garside normal form is a unique normal form and that thus two n-braid words α and β satisfy $\alpha \approx \beta$ if and only if $\alpha_G = \beta_G$. Executing this method in practise is time-consuming due to the size of the diagram. There exists an efficient polynomial-time algorithm to extract the maximal number of Δ_n from α' due to Jacquemard [44] which we shall not present here.

The word problem was first solved by Artin [7] and then by Garside [37]. Both of these solutions were algorithmic with exponential complexity. As mentioned above, Garside's algorithm can be made polynomial time due to a new algorithm by Jacquemard [44] to extract braids. The word problem for braids is important enough for many people to have studied it after Garside. The most efficient algorithm (linear in n and quadratic in L) is given in [15]. Below we present a new algorithm based on rewriting systems. It is not as efficient as the best algorithm for the word problem but it is easily generalizable to the conjugacy case and it is very simple to apply. Because of these features, we regard it as a competitive algorithm.

4.2 The Word Problem II: Rewriting Systems

In this section we will develop a new algorithm based on rewriting systems which are used as a tool in theoretical computer science. We begin with a finite alphabet of *constants* $\mathcal{A}$ and a finite set of *variables* $\mathcal{X}$. A *term* t is a finite ordered sequence of constants and variables $t = a_1 a_2 \cdots a_n$ with $n \geq 0$ (i.e. empty terms are allowed) and $a_i \in \mathcal{A} \cup \mathcal{X}$. A *word* w is a finite ordered sequence of constants $w = b_1 b_2 \cdots b_m$ with $m \geq 0$ and $b_i \in \mathcal{A}$. A *substitution* ρ for a term t is a map which assigns a word to each variable in t; the resultant word is denoted by ρt. A *term rewriting system* (TRS) $\mathcal{R} = \{(l_i, r_i)\}$ is a set of ordered pairs of terms l_i and r_i. Each ordered pair in $\mathcal{R}$ is referred to as a *rule* or *rewrite rule* and is often written in the form $l_i \to r_i$; the whole TRS is sometimes denoted by $\to_{\mathcal{R}}$. A TRS $\mathcal{R} = \{(l_i, r_i)\}$ is applied to a word w_0 by determining if w_0 contains the word ρl_i, for some substitution ρ, as a subword. If and only if w_0 contains ρl_i is ρl_i replaced by ρr_i. If $\to$ is a rewrite rule, then $\leftarrow$ is its inverse, $\leftrightarrow$ is its symmetric closure ($\leftarrow \cup \to$) and $\to^*$ is its reflexive-transitive closure ($\to \circ \to \circ \cdots \circ \to$). Two terms t and s are said to be *joinable* if there exists a term r such that $t \to^* r \leftarrow^* s$. Any l_i is called a *redex* and any r_i is called a *reduct* (these are abbreviations of reducible expression and reduced term).

A word w_0 is thus rewritten into a word w_1 if and only if $\mathcal{R}$ may be applied to w_0. We may generate a *rewrite chain* of words $w_0 \to_{\mathcal{R}} w_1 \to_{\mathcal{R}} \cdots$ in this manner. $\mathcal{R}$ *terminates* if and only if there exists no rewrite chain of infinite length. $\mathcal{R}$ is *locally confluent* if and only if any local divergence $\leftarrow \circ \to$ is contained in the joinability relation $\to^* \circ \leftarrow^*$. $\mathcal{R}$ is *confluent* if and only if any divergence $\leftarrow^* \circ \to^*$ is contained in the joinability relation $\to^* \circ \leftarrow^*$. $\mathcal{R}$ is *complete* if it is confluent and terminates. If $\mathcal{R}$ is complete a unique normal form exists for each word [8]; the final form obtained by applying $\mathcal{R}$ to the word a maximum number of times.

It should be noted that the computational power of term rewriting systems is identical to that of Turing machines, i.e. one may be simulated by the other [70]. According to the Church-Turing thesis [29], this means that any function which may reasonably be termed computable is computable using a TRS.

It was proven by Birkhoff [24] that the symmetric-reflexive-transitive closure $\leftrightarrow_{\mathcal{R}}^{*}$ of a TRS $\mathcal{R} = \{(l_i, r_i)\}$ is equivalent to the set of equations $\mathcal{E} = \{l_i = r_i\}$. It is an obvious corollary to Birkhoff's theorem that if there exists a complete TRS $\mathcal{R}$ over the alphabet $\mathcal{A} = \{f_i, f_i^{-1}\}$ for which $\leftrightarrow_{\mathcal{R}}^{*}$ contains exactly the equations $\{\mathcal{E}, f_i f_i^{-1} = e, f_i^{-1} f_i = e\}$, then $\mathcal{R}$ solves the word problem for the group $G = \langle \{f_i\}, E \rangle$. Note that $\mathcal{R}$ also solves the word problem for the monoid associated with G, i.e. the monoid obtained when the inverses of the generators are added to the set of generators and the fact that the generators and inverses are in fact inverses ($f_i f_i^{-1} = e, f_i^{-1} f_i = e$) added to the set of equations.

4.2.1 Termination

It is, in general, undecidable whether a TRS terminates or not [43]. Since any Turing machine can be modeled using a TRS this is essentially due to the undecidability of whether a Turing machine will stop, the Turing Halting Problem [71]. It is however decidable for a TRS without any variables [33]. Thus, in general, a termination proof is specific to a particular TRS and must be given for it. A common strategy for proving termination is to use a reduction order on the symbols involved in the TRS. We define a *reduction order* $<_o$ as a strict order over the alphabet and variables of the TRS which satisfies:

1. **compatibility**: For all terms u, v for which $u <_o v$, we have $xuy <_o xvy$ for any terms x and y.
2. **closure**: For all terms u, v for which $u <_o v$ and all substitutions σ, we have $\sigma u <_o \sigma v$.
3. **basis**: $<_o$ is well-founded, i.e. there exists a simplest term under $<_o$.

If one can show that every possible rewriting operation simplifies any term with respect to such an ordering, then the TRS terminates [32]. Furthermore, a TRS $\mathcal{R}$ terminates if and only if there exists a reduction order $<_o$ which satisfies $r_i <_o l_i$ for every rule $l_i \rightarrow r_i \in \mathcal{R}$ [8]. This is true because every step of the rewriting process simplifies the term and there exists a simplest term. Another useful result is that a TRS terminates if and only if it terminates for all instances of its redexes [34]. Some conditions under which the union of two terminating TRSs is terminating are analyzed in [34].

4.2.2 Confluence

Like termination, confluence is, in general, undecidable [8]. However, for terminating systems there exists a mechanizable method for deciding confluence [41] that rests on Newman's Lemma which states that a terminating TRS is confluent if and only if it is locally confluent [62] (we shall prove a generalization of this in lemma 4.11). Local confluence can be decided by a systematic method which searches for critical pairs in the TRS. The concept of critical pairs is difficult to trace in history; for an attempt at a historical survey see [27] and for a good technical treatment see [41].

Given a TRS $\mathcal{R} = \{(l_i, r_i)\}$, an *overlap* is a word $w = abc$ such that $ab = \rho l_i$ and $bc = \eta l_j$ for some words a, b and c, two (possibly equal) integers i and j and substitutions ρ and η. Clearly the overlap abc may be rewritten to both $\rho r_i c$ and $a \eta r_j$. An overlap is *non-critical* if the reducts are joinable, $\rho r_i c \leftrightarrow^*_{\mathcal{R}} a \eta r_j$ and *critical* otherwise. A *critical pair* is the (unordered) pair $(\rho r_i c, a \eta r_j)$ which arises from a critical overlap. It is obvious that if $\mathcal{R}$ contains critical pairs, it can not be confluent. The fact that the non-existence of critical pairs is both a necessary and sufficient condition for local confluence is called the Critical Pair Lemma [46]. Later we shall prove lemma 4.12 which contains the Critical Pair Lemma.

4.2.3 Completeness

If we can find a reduction order for a TRS $\mathcal{R}$, thereby prove termination and find that there are no critical pairs, $\mathcal{R}$ is complete and thus solves the word problem for $\leftrightarrow^*_{\mathcal{R}}$. A general procedure for what to do when we can not do this is called Knuth-Bendix completion from their seminal paper [50]. Again a historical account of this procedure is tangled and [27] is an attempt to unravel it. We shall follow the common practice to call it Knuth-Bendix completion even though, by their own admittion, the initial idea was not theirs.

Suppose we have a set of equations $\mathcal{E}$ on an alphabet $\mathcal{A}$ and a total order $<_{\mathcal{A}}$ (this is a reduction order) on $\mathcal{A}$. Construct a TRS $\mathcal{R}$ from $\mathcal{E}$ by creating a rule $l_i \rightarrow r_i$ in $\mathcal{R}$ from the equation $l_i = r_i$ in $\mathcal{E}$ for all equations in $\mathcal{E}$ such that the rules are ordered such that $l_i >_{\mathcal{A}} r_i$. Now $\leftrightarrow^*_{\mathcal{R}}$ is equivalent to $\mathcal{E}$ and each rule represents a simplification in terms of $<_{\mathcal{A}}$. Clearly there exists a simplest word, the empty word, and so $\mathcal{R}$ terminates.

If there are no critical pairs, $\mathcal{R}$ is locally confluent and thus complete. If there are critical pairs, order them with respect to $<_{\mathcal{A}}$ and append them to $\mathcal{R}$ as new rules. Termination still holds and so we continue this process. We may delete duplicate or redundant rules from $\mathcal{R}$ between the steps of this method to obtain a smaller TRS. If this method terminates, we have a complete system [50] [42]. If it does not terminate, a complete system may still exist which contains an infinite number of rules. It is possible to collect an infinite number of rules into a finite number of rules by introducing variables. The Knuth-Bendix algorithm has been implemented by several people and can be used to determine, in some cases, whether a complete system exists. The CiME implementation was used for this thesis [53]. Producing rules with variables and proving the non-existence of critical pairs is, at present, beyond the computer implementations and must be done manually.

The process described here is simplified; there are more pitfalls, in general, and the method has been considerably extended to take into account many other features (many relevant references are in [27]). The method as described is enough for our purposes here however and is generally enough for a word problem in a finitely presented group.

Having reviewed TRS's, we are now in a position to find a TRS for the word problem in B_n. The braid group B_n is defined formally as

$$B_n = \langle \{\sigma_1, \sigma_2, \cdots, \sigma_{n-1}\} : \sigma_i \sigma_{i+1} \sigma_i = \sigma_{i+1} \sigma_i \sigma_{i+1};$$
$$\sigma_i \sigma_j = \sigma_j \sigma_i \text{ for } |i - j| > 1 \rangle \tag{51}$$

Given a finitely presented group $G = \langle X, E \rangle$, we can define an associated monoid $M(G) = \langle X \cup X^{-1}, E \cup aa^{-1} = 1 \rangle$ for any $a \in X$. It is clear that the equivalence and conjugacy classes of the group G and the monoid $M(G)$ are identical. In order to solve the word problem for B_n, we augment the monoid $M(B_n)$ with the generator of the center of B_n, Δ_n^2 to form the monoid

$$M^+(B_n) = \langle \{\sigma_1^{\pm 1}, \sigma_2^{\pm 1}, \cdots, \sigma_{n-1}^{\pm 1}, \Delta_n^{\pm 2}\} : \Delta_n^{\pm 2} \sigma_i = \sigma_i \Delta_n^{\pm 2};$$
$$\Delta_n^{\pm 2} \Delta_n^{\mp 2} = \sigma_i^{\pm 1} \sigma_i^{\mp 1} = e;$$
$$\sigma_i^{\pm 1} \sigma_j^{\pm 1/\mp 1} = \sigma_j^{\pm 1/\mp 1} \sigma_i^{\pm 1} \text{ for } |i - j| > 1;$$
$$\sigma_i^{\pm 1} \sigma_{i+1}^{\pm 1} \sigma_i^{\pm 1} = \sigma_{i+1}^{\pm 1} \sigma_i^{\pm 1} \sigma_{i+1}^{\pm 1} \rangle \tag{52}$$

It is obvious from the definition of the monoid $M^+(B_n)$ that a solution of its word and conjugacy problems provides a solution for the word and conjugacy problems in the group B_n.

We will use Knuth-Bendix completion upon the oriented rules of $M^+(B_n)$ under the reduction order $<_b$

$$\Delta_n^2 <_b \Delta_n^{-2} <_b \sigma_1 <_b \sigma_2 <_b \cdots <_b \sigma_{n-1} <_b \sigma_1^{-1} <_b \sigma_2^{-1} <_b \cdots <_b \sigma_{n-1}^{-1} \tag{53}$$

In practice, this process is laborious and would occupy prodigious space if described in detail. For this reason, we will simply state the result and prove it to be correct.

For what follows, we shall represent a braid of the form $\Delta_n^{2k} P$ as the pair (k, P). The reason for this is to effectively remove from the braid, in the process of rewriting, any subbraid which lies in the center of the braid group B_n. The reason for this will become apparent when we extend our solution to the conjugacy problem. Removing any Δ_n^{2k} from any part of a braid can be done without loss of information because Δ_n^{2k} is the generator of the center of B_n and thus its position is irrelevant. By Knuth-Bendix completion and the necessary manual labor, we obtain the following rewriting system.

$$
\begin{aligned}
\mathcal{W}_n = \{ &(1)\ \sigma_i^{-1} \to \prod_{j=1}^{i-1} [d_{j,1}a_{1,j}]\, d_{i,1}a_{1,i-1} \prod_{j=i+1}^{n-1} [d_{j,1}a_{1,j}]\ \ \&\ \ k \to k-1; \\
&(2)\ \sigma_i\sigma_j \to \sigma_j\sigma_i \text{ for } j < i-1; \\
&(3)\ \sigma_i\sigma_{i-1}P\sigma_i \to \sigma_{i-1}\sigma_i\sigma_{i-1}P; \\
&(4)\ \sigma_i\sigma_{i-1}Q\sigma_{i-1}Rd_{i,j} \to \sigma_{i-1}\sigma_i\sigma_{i-1}Qd_{i-1,j}\sigma_iR^+ \text{ for } j < i; \\
&(5)\ \prod_{i=1}^{n-1} d_{i,1}a_{1,i}S_i \to \prod_{i=1}^{n-1} S_i\ \ \&\ \ k \to k+1\ \}
\end{aligned}
$$

$$(54)$$

The variables P, Q, R and S_i are (possibly empty) words in the generators σ_k (and *not* their inverses σ_k^{-1}) subject to the restriction that the highest generator index k is $i-2$, $i-2$, $i-1$ and i respectively and the lowest generator index in R is j, where i and j refer to the values of the generator indices of the respective rules. The word R^+ is obtained from R by increasing all generator indices in R by one. Note that rules 1 and 5 require two replacements to be made simultaneously. A similar, unpublished TRS was also found using Knuth-Bendix completion by Yoder [78]. Rules 1 and 5 are simple to understand; the other rules are illustrated in figure 29.

Theorem 4.4. $\mathcal{W}_n$ *is complete and solves the word problem for* B_n.

Proof. It can be checked easily but laboriously that the system terminates, there are no critical pairs and that its symmetric-transitive closure is the monoid we began with which proves the theorem. $\square$

The rules of a TRS are to be applied in a non-deterministic way and a complete TRS always reaches the unique normal form no matter what strategy of rule application is chosen [8]. Since $\mathcal{W}_n$ is complete and all strategies are equivalent, we will choose the following strategy.

Algorithm 4.5 *Input: A word* $w \in B_n$. *Output: A word* $w' \in B_n$ *which is the unique representative of the equivalence class of* w.

1. Apply rule 1 of $\mathcal{W}_n$ as many times as possible.
2. Apply rules 2, 3, 4 and 5 of $\mathcal{W}_n$ as many times as possible in order proceeding to the next rule only if the current can no longer be applied.
3. Loop step 2 until no rule may be applied to the word at all. In this case w' has been found.

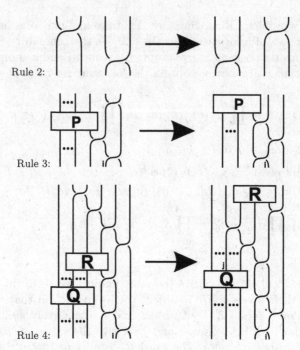

Fig. 29 Rules 2, 3 and 4 of TRS $\mathcal{W}_n$ illustrated.

It is clear that algorithm 4.5 solves the word problem from the completeness of $\mathcal{W}_n$ and the fact that once rule 1 is applied as many times as possible, it can not be applied again no matter what other rewrite steps follow as there will be no more inverse generators. From this algorithm, we are able to deduce the computational complexity of this word problem solution.

Theorem 4.6. $\mathcal{W}_n$ *solves the word problem for any word* $w \in B_n$ *of word length* l *with complexity* $O\left(l^2 n^4\right)$.

Proof. This can be checked easily by counting how many times the rules are used. $\square$

4.3 The Conjugacy Problem I: Garside's Solution

Extending the word problem solution of Garside to the conjugacy case is not hard. First we define two new terms.

Definition 4.7. If a positive word $b = if$ for two braid words i and f such that $1 \leq L(i) \leq L(b)$, then i is called an *initial route* of b and f an *associated final route*. Note that the definition requires exact equality $(=)$ and not group theoretic equivalence $(\approx)$.

We begin by constructing all possible initial routes of Δ_n. This is easily done using $D(\Delta_n)$. Replace all of these routes by the base of their diagrams and delete duplicates. This is the *set of initial routes* of Δ_n and denoted by $I(\Delta_n)$.

We will now form a set $S(\beta)$ for a braid word β called the *summit set* of the braid. This set is the analogue of the Cayley diagram, which we used for the word problem, for the conjugacy problem. We construct the summit set as follows. First construct the Garside normal form of β, i.e. β_G, and note its exponent. The set $S_1(\beta)$ is then constructed by conjugating β_G by each word of $I(\Delta_n)$, computing the Garside normal form of the result and deleting all those words of lower exponent than β_G. Iteratively, the set $S_i(\beta)$ is obtained from the set $S_{i-1}(\beta)$ by conjugating each element of $S_{i-1}(\beta)$ by each word of $I(\Delta_n)$, computing the Garside normal form of the result and deleting all those words of lower exponent than β_G.

Suppose the Garside exponent of β_G is m and its exponent sum is s. Every word in any $S_i(\beta)$ has exponent sum s and Garside exponent $p \geq m$ by definition. The Garside remainder of β_G is $\underline{\beta_G}$ and so we have

$$s = L\left(\underline{\beta_G}\right) - mL\left(\Delta_n\right) \tag{55}$$

$$m = \frac{L\left(\underline{\beta_G}\right) - s}{L\left(\Delta_n\right)} \tag{56}$$

$$p \leq \frac{s - L\left(\underline{\beta_G}\right)}{L\left(\Delta_n\right)} \tag{57}$$

but we also have $p \geq m$ and so there are only finitely many p that satisfy the requirements. Hence there are only finitely many distinct words with Garside exponent p and exponent sum s. Thus the process of constructing the sets $S_i(\beta)$ terminates, i.e. there exists a finite integer j such that $S_j(\beta) = S_{j+1}(\beta)$.

Definition 4.8. We define the summit set $S(\beta)$ to consist of all the elements of $S_j(\beta)$ which have maximal Garside exponent. The *summit exponent t* is the Garside exponent of all these braid words and the *summit remainder* is the Garside remainder $\underline{\beta_S}$ of the unique element in $S(\beta)$ with smallest tail. The *summit form* of β is then $\beta_S = \Delta_n^t \underline{\beta_S}$.

We state, without proof, the main result of Garside.

Theorem 4.9 (Garside's Conjugacy Theorem [37]). *Two braid words $\beta_1, \beta_2 \in B_n$ are conjugate if and only if their summit forms are identical.*

Proof. For a proof, see [37, 14]. $\square$

Like the word problem, the conjugacy problem is very important for braid and knot theory and so has received much attention. The first solution was produced by Garside [37] and the best algorithms are given in [15, 36]. We should remark that Jacquemard [44] has used his extraction algorithm to obtain good practical results for small n. All these algorithms still require an exponential amount of computation time as a function of n and L. The important question of whether conjugacy is solvable in polynomial time is solved positively in the next section.

4.4 The Conjugacy Problem II: Rewriting Systems

4.4.1 Conjugacy in Free Groups

Suppose that $G = \mathcal{F}_n$ the free group of rank n. This group is generated by n elements $\{f_i\}$ for $1 \leq i \leq n$ and no relations [45]. A general word $w \in \mathcal{F}_n$ takes the form

$$w = f_{s_1}^{p_1} f_{s_2}^{p_2} \cdots f_{s_m}^{p_m}, \qquad 1 \leq s_k \leq n \tag{58}$$

Since there are no relations in $\mathcal{F}_n$, the word w is unique over its equivalence class if and only if $s_i \neq s_{i+1}$ for all i. This condition is trivially obtained from any word $w \in \mathcal{F}_n$ by applying the (obviously) complete rewriting system

$$\mathcal{R}_w\left(\mathcal{F}_n\right)_i = \{f_s^p f_s^q \to f_s^{p+q}, \forall 1 \leq s \leq n\} \tag{59}$$

Thus $\mathcal{R}_w\left(\mathcal{F}_n\right)$ solves the word problem in any free group $\mathcal{F}_n$. Moreover, it does so in a time proportional to the length of the word w.

Consider now the conjugacy problem in $\mathcal{F}_n$. We define the ith *cyclic permutation* $C^i(w)$ of a word w in the general form of equation (58) by

$$C^i(w) = f_{s_j}^{p_j'} \cdots f_{s_{m-1}}^{p_{m-1}} f_{s_m}^{p_m} f_{s_1}^{p_1} f_{s_2}^{p_2} \cdots f_{s_j}^{p_j''} \tag{60}$$

such that

$$p_j' + \sum_{k=j+1}^{m} p_k = i \tag{61}$$

Intuitively, the ith cyclic permutation is obtained by taking the last i generators in the word w and moving them to the front of the word w one by one. We shall say that two words w and w' are *cyclicly permutable* (denoted $\approx_{cp}$) if and only if there exist an i such that $C^i(w) \approx w'$. It is obvious that cyclic permutability forms an equivalence relation for any group G.

Proposition 4.10. *For any group G, the equivalence relation of cyclic permutability ($\approx_{cp}$) is identical to that of conjugacy ($\approx_c$).*

Proof. Any group G has a presentation which may be obtained from some free group $\mathcal{F}_n$ of rank n by adding relations [31]. Moreover, if the conjugacy problem is solvable in one representation, it is solvable in all [57]. Suppose $w \approx_{cp} w'$, then there exists an i for which

$$w' \approx C^i(w) = f_{s_j}^{p'_j} \cdots f_{s_{m-1}}^{p_{m-1}} f_{s_m}^{p_m} f_{s_1}^{p_1} f_{s_2}^{p_2} \cdots f_{s_j}^{p''_j} \tag{62}$$

where

$$p'_j + \sum_{k=j+1}^{m} p_k = i \tag{63}$$

Let

$$\gamma = f_{s_1}^{p_1} f_{s_2}^{p_2} \cdots f_{s_j}^{p''_j} \tag{64}$$

Then

$$w' \approx f_{s_j}^{p'_j} \cdots f_{s_{m-1}}^{p_{m-1}} f_{s_m}^{p_m} \gamma \tag{65}$$

$$\approx \gamma^{-1} \gamma f_{s_j}^{p'_j} \cdots f_{s_{m-1}}^{p_{m-1}} f_{s_m}^{p_m} \gamma \tag{66}$$

$$\approx \gamma^{-1} w \gamma \tag{67}$$

Thus we have $w \approx_c w'$. Now suppose $w \approx_c w'$, then there exists a γ such that

$$w' \approx \gamma^{-1} w \gamma \tag{68}$$

If the word length of γ is $L(\gamma)$, then we have

$$C^{L(\gamma)}(w') \approx \gamma \gamma^{-1} w \approx w \tag{69}$$

Thus $w \approx_{cp} w'$. $\qquad\square$

We will refer to the set of words which contains the word w and all its cyclic permutations as the *cyclic word* $c(w)$. If $L(w) = m$, then this set contains $|c(w)| = m$ elements. Given two cyclic words $c(w)$ and $c(w')$ we test their equivalence by attempting to construct an isomorphism $\iota : c(w) \to c(w')$ such that $\iota(a) = a$ for all $a \in c(w)$. Clearly $|c(w)| = |c(w')|$ is a necessary condition for the existence of ι. If and only if ι exists, the cyclic words are considered equal, $c(w) = c(w')$. If and only if $c(w) = c(w')$, we have $w \approx_c w'$ by proposition 4.10. The set $c(w)$ may be visualized as the word w "made circular" as in figure 30.

The existence of ι may be established by testing the members of $c(w)$ for equality with the members of $c(w')$ pairwise in the following manner: Select from $c(w)$ an arbitrary member, a say. Check a for equivalence with all members of $c(w')$. Clearly, if and only if there exists a $b \in c(w')$ such

Fig. 30 The word w given in equation (58) bent into a circle. While the circularity removes the notions of beginning and end of a word, it preserves the directionality of it.

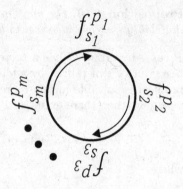

that $a = b$, an ι exists. Since every word has length m and there are m words in $c(w')$, this comparison will take a time proportional to m^2. Thus it is possible to test the equivalence of two cyclic words of length m with complexity $O\left(m^2\right)$.

4.4.2 Rewriting Systems for Cyclic Words

We shall call a TRS *cyclicly terminating, cyclicly confluent* and *cyclicly locally confluent* if it is respectively terminating, confluent and locally confluent under application to all cyclic words over the alphabet of the TRS. It is obvious from the above discussion that a cyclicly complete TRS solves the conjugacy problem. For this reason it is important to develop results about cyclic completeness along the lines of the results for linear words in order to obtain a conjugacy solution.

4.4.3 Termination in Cyclic Rewriting Systems

We have seen that a TRS $\mathcal{R}$ terminates if and only if a reduction order exists [8]. In what follows, we shall assume that this reduction order is a total order; note that this is a stricter requirement than that of a reduction order. Suppose that the alphabet of $\mathcal{R}$ is $\mathcal{A} = \{f_i\}$ for $1 \leq i \leq p$. By assumption, p is finite. Consider the total order $<_\mathcal{R}$ defined by $f_i <_\mathcal{R} f_{i+1}$ for all i. This can be done without loss of generality as a mapping from $\mathcal{A}$ to itself can change the order. Recall that $\mathcal{R}$ terminates if and only if $r_i <_\mathcal{R} l_i$ for every rule $l_i \to r_i$ $\in \mathcal{R}$.

We introduce an integer valued *weight* metric function $g(w)$ and an integer valued *length* metric function $L(w)$ on the set of words w written on the alphabet $\mathcal{A}$. The metrics satisfy

$$g\left(f_{a_1}f_{a_2}\cdots f_{a_m}\right) = g\left(f_{a_1}\right) + g\left(f_{a_2}\right) + \cdots + g\left(f_{a_m}\right) \tag{70}$$

$$L\left(f_{a_1}f_{a_2}\cdots f_{a_m}\right) = L\left(f_{a_1}\right) + L\left(f_{a_2}\right) + \cdots + L\left(f_{a_m}\right) \tag{71}$$

$$L\left(f_i\right) = 1 \tag{72}$$

$$g\left(f_i\right) < g\left(f_{i+1}\right) \tag{73}$$

We shall call a rule *length reducing* if $L(r_i) < L(l_i)$ and *weight reducing* if $g(r_i) < g(l_i)$. Any rule is a *c-obstruction* (for commutation-obstruction) if and only if it keeps constant both length and weight. That is, it is a rule which changes the position of the letters only.

A c-obstruction obstructs cyclic termination as there exist cyclic words which would give rise to an infinite rewriting chain due the changing of relative position of subwords by the c-obstruction. An example is the cyclic word $c(\alpha\beta\alpha\beta)$ under the TRS $\mathcal{R} = \{\alpha\beta \rightarrow \beta\alpha\}$. The rewriting chain will loop between the two states $c(\alpha\beta\alpha\beta)$ and $c(\beta\alpha\alpha\beta)$; the period of the loop may, in general, be arbitrarily large. Such looping may be dealt with in two ways. Firstly, one may compare each new cyclic word with the entirety of the rewrite chain so far enumerated. If equality is found, looping has been detected and one may stop. Secondly, one may determine if a subword of the current word commutes with the rest of the word. If this can be determined and such a subword is found, the subword may be extracted from the word and the two words should then be rewritten separately. The first method is computationally expensive and does not produce a unique normal form as we would have to consider the entire loop at the end of the rewrite chain as the identifying set of the word. The second method is not necessarily applicable but if it is, it will terminate in a set of subwords which uniquely identify the word. The advantage of the second method over the first is that the number of elements in the set has an upper bound.

The braid groups, as we have seen, have non-trivial center. In fact the generator of the center contains every generator. This fact makes it possible to rewrite any inverse generator in terms of inverses of the generator of the center multiplied by generators. As in the word problem case, we shall remove the generators of the center from the word and thus this replacement rewrites the entire braid word with which we shall work in terms of generators only. For the splitting of words to get rid of c-obstructions to fail we need to be in a situation in which we have a word abc such that $abc \approx cab$, $ab \approx ba$, $ac \not\approx ca$, $bc \not\approx cb$. This can only occur if there is cancellation between a and b in the word ab but this can not happen if there are no inverse generators. This proves that in groups in which inverse generators may be replaced by inverses of elements of the center and generators, this method of overcoming c-obstructions is valid.

We conjecture that a TRS $\mathcal{R}$ cyclicly terminates if and only if it terminates and contains no c-obstructions or contains c-obstructions that can be removed in the above way.

4.4.4 Confluence in Cyclic Rewriting Systems

Newman's Lemma [62] extends easily to the cyclic case as we show below.

Lemma 4.11. *A cyclically terminating TRS $\mathcal{R}$ is cyclicly confluent if and only if it is cyclicly locally confluent.*

Proof. The proof is similar to the standard proof (see [41]) and follows immediately from figure 31.   □

The Critical Pair Lemma states that a TRS is locally confluent if and only if it has no critical pairs. Recall that a critical pair arises from an overlap of two redexes in a word which gives rise to a local divergence of rewriting paths which do not meet again. Given a TRS $\mathcal{R} = \{(l_i, r_i)\}$, a *cyclic overlap* is a cyclic word $c(w) = c(abcd)$ such that $abc = \rho l_i$ and $cda = \eta l_j$ for some words a, b, c and d, two (possibly equal) integers i and j and substitutions ρ and η. The cyclic overlap $c(abcd)$ is rewritten to both $c(\rho r_i d)$ and $c(b\eta r_j)$. A cyclic overlap is *non-critical* if the reducts are joinable, $c(\rho r_i d) \leftrightarrow^*_{\mathcal{R}} c(b\eta r_j)$ and *critical* otherwise. A *cyclic critical pair* is the (unordered) pair of cyclic words $(c(\rho r_i d), c(b\eta r_j))$ which arises from a cyclic critical overlap. It is obvious that if $\mathcal{R}$ contains cyclic critical pairs, it can not be cyclicly confluent.

For example, consider the rewrite system $\mathcal{R} = \{abxba \to cxc\}$ over the alphabet $\mathcal{A} = \{a, b, c\}$ and some variables x and y. Clearly $\mathcal{R}$ contains the cyclic critical overlap $abxbabyb$ which is to be rewritten into $bxbcyc$ and $cxcbyb$. This cyclic critical pair may be resolved by noting that if the variable contained between the c letters is less than the other, it is that cyclic word which is to be prefered under the lexicographic order $c < b < a$. That is, we have to add a conditional rule depending on the relative value of the variables. This global rule must be applied, if applicable, with preference over the ordinary local rule. In this way we have extended Knuth-Bendix completion to the cyclic case; note that all rules added in this procedure are *global* whereas the usual rules of normal TRS's are *local*. We shall now prove the extention of the Critical Pair Lemma for the cyclic case.

Fig. 31 The proof of Newman's Lemma (lemma 4.11) in diagrammatic form. We begin at the top with a local divergence which is rectifiable by assumption and thus by induction any global divergence is also rectifiable. It is because of this diagrammatic proof that Newman's Lemma is also known as the Diamond Lemma.

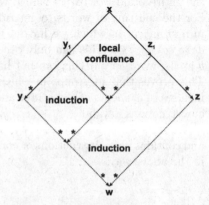

Lemma 4.12. *A TRS $\mathcal{R} = \{(l_i, r_i)\}$ is cyclicly locally confluent if and only if it contains neither critical nor cyclicly critical pairs.*

Proof. This can easily be checked by going through all possible types of overlap.   □

It should be emphasized that the proof lemma 4.12 does not make any assumptions about the termination of $\mathcal{R}$. So we have a definite method for attempting to find a conjugacy problem solution in terms of rewriting. We shall use the braid groups to give an example of this completion process.

4.5 Markov's Theorem

Recall that Markov's theorem says

Theorem 4.13 (Markov). *Two braids $\alpha \in B_n$ and $\beta \in B_m$ have isotopic closures if and only if α can be transformed into β by a finite number of applications of conjugacy and stabilization moves.*

Corollary 4.14. *The closure of the braid $\alpha \in B_n$ is isotopic to the unlink of k components if and only if α can be transformed into the trivial braid in B_k by using conjugacy and stabilization moves.*

Conjugacy was discussed at length already. Stabilization is the move $\alpha \leftrightarrow \alpha\sigma_n$ with $\alpha \in B_n$ (see figure 32). While the conjugacy move is a move within a particular braid group, the stabilization move connects two adjacent braid groups. Therefore the question of detecting closed braid equivalence turns into a combinatorial question about the infinite family of braid groups.

Given a knot, we may produce an equivalent knot by taking any segment and twisting it about an axis in the projection plane by π while keeping the rest of the knot stationary. This procedure corresponds to the zeroth Reidemeister move and adds one crossing to the diagram. Any crossing of

Fig. 32 Both conjugacy and stabilization are displayed here. We begin with braid B. Conjugation surrounds B with A and A^{-1} on opposite sides which clearly cancel due to the closure. Stabilization introduces a simple loop at the bottom right of the braid, adds a new string to the braid and thus increases the braid group index by one.

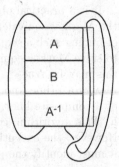

this type is called *nugatory*. If we represent a knot by a closed braid by
virtue of Alexander's theorem, we may also add such nugatory crossings via
a combinatorial move, called the *Markov* or *stabilization* move. Stabilizing a
braid $\alpha \in B_n$ corresponds to the operation $\alpha \to \alpha \sigma_n^{\pm 1}$ or its inverse. Clearly
stabilization increases or decreases the number of strings in the braid and so
represents a move in the family of braid groups as opposed to the conjugacy
and equivalence moves which are contained in a single braid group.

Markov stated in 1935 [55] that two closed braids are topologically equivalent if and only if they differ by stabilization and conjugacy moves (recall that
conjugacy contains equivalence). This statement became known as Markov's
theorem and was first proven in [14]. In its original form, Markov's theorem
assumes that the closed braid is embedded in S^3 or R^3, this can however, be
generalized to an arbitrary 3-manifold [51]. Markov's theorem transforms the
link isotopy problem to a combinatorial question about braids. If two braids
$\alpha \in B_n$ and $\beta \in B_m$ (with n and m possibly different) are related by stabilization and conjugacy, they are called *Markov equivalent* which is denoted by
$\approx_M$. The decision problem of whether $\alpha \approx_M \beta$ is called the *Markov problem*
or the *algebraic link problem*. It is possible to find a single move of which
both stabilization and conjugacy are special cases and to formulate, in this
way, Markov equivalence in terms of this so called L-move [52]. While this
L-move is intuitive, it is not obvious whether the problem has been simplified
by this reformulation.

The first question which arises is whether there exist non-conjugate
Markov equivalent braid words in the same braid group, that is whether
a solution to the conjugacy problem will solve the Markov problem. This
is negatively resolved by showing that the two 4-braids $\alpha = \sigma_1^m \sigma_2^n \sigma_1^p$ and
$\beta = \sigma_1^m \sigma_2^p \sigma_1^n$ with m, n, p different, odd and at least three in absolute value
are not conjugate but Markov equivalent [61]. It might be thought that it
should be possible to reduce the number of strings in a closed braid equivalent of the unknot to one. This is true as all equivalent closed braids can be
reached from each other via Markov's theorem but the transition involves,
in general, increasing the number of strings before they may be reduced to
a single string. In other words, a greedy reduction of strings does not reach
the minimum string number, also known as the braid index (not even for the
unknot representatives) [58].

It is a practical observation that finding a series of moves to demonstrate
the Markov equivalence of two closed braids is very difficult. The difficulty
of finding such a sequence has lead Birman to believe that it may be simpler
to solve Markov equivalence for two braids representing prime knots. While
this may be true, it is not, in general, easy to decide whether a braid represents a prime knot. Schubert [67] proved that the factorization sequence
of a composite knot is unique and has found an algorithm [68] which finds
it. This algorithm, consequently, is able to decide whether a knot is prime.
However, the execution of the algorithm rests on Hemion's algorithm since
it must identify the prime factors of the knot, thus no longer necessitating a

solution of the Markov problem since it already solves the link isotopy problem (albeit impractically so). This also shows that this method of deciding primality is not practical. Birman conjectures that a braid represents a prime knot if and only if it is not conjugate to a split braid.

Furthermore, if Birman's conjecture is true and we were to find an algorithm to decide whether a braid was conjugate to a split braid, we would have to solve the Markov problem for this restricted class of braids. If this could be done, we would have a solution to the Markov problem since every braid could be decomposed into its split components and pair-wise tested for non-split Markov equivalence. This would not only resolve isotopy but also give the unique prime knot factorization of the knots. Birman's conjecture is unproven and there exists no algorithm to test whether a braid is conjugate to a split braid. It is possible, however, to solve the Markov problem for certain quotient groups of the braid groups [23].

Since the word and conjugacy problems are contained in the Markov problem, solutions for these are desirable and have been given numerous times as mentioned before. The stabilization move represents the final hurdle before link isotopy is algorithmically decidable and thus it would be interesting to know when a braid $\alpha \in B_{n+1}$ is conjugate to a braid $\gamma \sigma_n^{\pm 1}$ where γ contains only the generators σ_i for $1 \leq i \leq n - 1$, for then one could reduce α to γ using the Markov move. While this has been done [56], the algorithm depends on Garside's conjugacy algorithm [37] which has exponential complexity. Moreover, if two braids were reduced in this way to the minimum string number, they are not, in general, conjugate in this final braid group if they are Markov equivalent and thus this decision procedure does not solve the Markov problem either.

We have defined the exponent sum $exp(\alpha)$ of a braid α as the sum of the exponents of the Artin generators of α. It is obvious that the exponent sum is a conjugacy class invariant but not a Markov class invariant because of stabilization. Thus it is possible for two braids to be Markov equivalent and have different exponent sums. In getting from one braid to the other, the exponent sum must be made equal somewhere in the chain of moves; this can clearly only be accomplished using stabilization. Stabilization can increase or decrease the exponent sum depending whether we add σ_n or σ_n^{-1} or remove either of these. It also changes the number of strings. We may think that starting from a positive braid, we should be able to reach any Markov equivalent positive braid by going through a pure positive sequence of braids; that is, we may think that positive Markov equal braids are positively Markov equal. We note that this would only be possible if the difference in exponent sum between the two braids was precisely their difference in number of strings. We conjecture that positive Markov equal braids are not positively Markov equal.

Much work was done by Birman and Menasco on various properties of links which could be determined from their closed braid representatives (this work was published in the six-paper series [16], [17], [18], [19], [20] and [21]). They prove that there exists a complete numerical invariant for knots but find this invariant only for knots which are closed 3-braids. The invariant for closed 3-braids is described extensively and can be used to determine the braid index and whether the knot is split, composite, amphicheiral or invertible. They also define a new type of move on braids, the exchange move, and prove a Markov-like theorem for it. See [22] for a summary of this work.

4.6 The Minimal Word Problem

A well-known problem of combinatorial braid theory is the minimization problem: Given a braid $A \in B_n$ find a braid A_m such that $A \approx A_m$ and $L(A_m) \leq L(A^*)$ for any braid $A^* \approx A$ where $L(A)$ denotes the word length of the braid A.

In the Artin representation of B_n, the number of generators required to write down a braid word, its length, is equal to the crossing number of the topological braid. In practice, we find that by moving a few of the strings of the topological braid, its crossing number may be reduced, making the braid simpler. It would be especially useful to possess a general method to compute an equivalent braid of minimum crossing number. Apart from many applications, this problem is well-known in combinatorial braid theory and is of independent mathematical interest.

Given a braid $A \in B_n$ in the Artin generators, the question whether there exists an equivalent braid $A' \in B_n$ of shorter length has been shown to be NP-Complete by Paterson and Razborov [63]. Not only does this mean that this question is computationally equivalent to all other NP-Complete problems, it also means that (unless P = NP) any algorithm which answers the question would execute in exponential-time in n. Since Paterson and Razborov's result refers to the minimization problem for general n, we ask whether it is also an NP-complete question for particular n. This question is explicitly asked as open question 9.5.6 on page 209 of [36] and it seems to have been negatively answered in an unpublished preprint by Tatsuoka five years earlier but we were unable to obtain it [69].

In proving the NP-Completeness of the problem, Paterson and Razborov showed that the problem can be reduced to a known NP-Complete problem. This does not however provide a usable algorithm. For 3-braids, a linear complexity algorithm has been found [11] but no general algorithm for $n > 3$ exists. A minimization algorithm in the band-generator presentation of the braids groups has been found for $n = 3, 4$ but the length of the braid in this presentation is not equal to the crossing number [76] [47]. It is untypical of a group for which the word problem is solvable that no unique normal form of minimal length in some naturally arising presentation exists for the braid groups. A unique normal form of minimal length in certain natural presentations of free groups, HNN-extensions and free products exists, for example.

After a little experimentation, it is clear that a braid must, in general, be increased in length before it may be reduced to minimum length algebraically. We show that a certain readily obtained braid provides an upper bound for this necessary increase in length and prove several properties of this braid. We explicitly construct a set of words which must be searched for a certain property in order to obtain a minimal length representative of any braid. This constitutes an algorithm to solve the minimization problem. Since the set of words which must be searched is, in the worst case, exponential in size, the algorithm takes an exponential amount of time to complete.

Exercise 4.15. Find a braid which is non-minimal in length and which must be increased in length (by introducing pairs like $\sigma_i \sigma_i^{-1}$) before it can be shortened to minimal length. An example is the braid $\sigma_2 \sigma_1^{-2} \sigma_2^{-2} \sigma_1 \sigma_2^{-1} \sigma_1 \sigma_2^2 \sigma_1 \sigma_2$.

Denote by A_m any braid which satisfies $A_m \approx A$ and $L(A_m) \leq L(A^*)$ for all braids $A^* \approx A$. We now prove a basic lemma which connects A_{max} and A_m. Recall that $A_{max} = \Delta_n^{-s(A)} A'$ where $s(A)$ is the number of inverse generators in A and A' is positive.

Theorem 4.16. *For any braid A, it is possible to obtain A_m from A_{max} by operations which monotonically decrease or keep constant the length of the braid.*

Proof. By construction $A_m \approx A_{max} \approx A$ and A_{max} and A are at least as long A_m. Exponent sum is an equivalence class invariant so that $s(A_m) \leq s(A)$. Replace each inverse generator in A_m with the braid given in proposition 4.1 and then use equation $\sigma_i \Delta_n = \Delta_n \sigma_{n-i}$ to bring all the fundamental braids to the front to obtain the braid

$$A_{m_{max}} = \Delta_n^{-s(A_m)} A_m' \tag{74}$$

$$\approx \Delta_n^{-s(A_m)} \Delta_n^{s(A_m)-s(A)} \Delta_n^{s(A)-s(A_m)} A_m' \tag{75}$$

$$\approx \Delta_n^{-s(A)} \Delta_n^{s(A)-s(A_m)} A_m' \tag{76}$$

But $A_{max} = \Delta_n^{-s(A)} A'$ and since the braid groups are left-cancelative [37], we have that

$$\Delta_n^{s(A)-s(A_m)} A_m' \approx A' \tag{77}$$

with both words positive. Since positive words are positively equal [37], there exists a sequence of braids B_i for $0 \leq i \leq q$ with $B_0 = A'$, $B_q = \Delta_n^{s(A)-s(A_m)} A_m'$, B_j and B_{j+1} different by a single application of the braid group's defining relations and B_i positive for all i. Since exponent sum is an equivalence class invariant, $L(B_i) = L(A')$ for all i.

From A_{max} we may thus reach the form of $A_{m_{max}}$ in equation (76) keeping the length of the braid constant. From this form, we may reach A_m by operations which monotonically decrease or keep constant the length of the braid.

Thus there exists a sequence of braids W_i for $1 \leq i \leq p$ with $W_0 = A_{max}$, $W_p = A_m$, W_j and W_{j+1} different by a single application of the braid group's defining relations and $L(W_{j+1}) \leq L(W_j)$, which proves the lemma. □

Theorem 4.16 basically establishes that we may reach a minimum length representative from A_{max} by rearranging and cancelling generators only; it thus, in principle, removes the difficulty we pointed out in the introduction of occasionally having to increase the length before being able to decrease it to an absolute minimum.

We present an extension to the Cayley diagram construction which draws the diagram of any braid word (as opposed to positive braid words only). The diagram is a list of all those braid words which may be obtained from the given word by rearranging only.

Algorithm 4.17 *Input: A braid word A. Output: A list $D(A)$ of all braid words B which may be obtained from A by rearranging of generators only.*

1. Define the diagram of zeroth order as the set $D_0(A) = \{A\}$.
2. The set $D_i(A)$ is obtained from the set $D_{i-1}(A)$ by the following procedure:

   a. Fix attention on a particular member α of $D_{i-1}(A)$. We read α from left to right and decide at each position whether we may apply any of the moves in equations (78) to (81).

   $$\sigma_i \sigma_j \leftrightarrow \sigma_j \sigma_i \text{ for } |i - j| > 1 \tag{78}$$
   $$\sigma_i \sigma_{i+1} \sigma_i \leftrightarrow \sigma_{i+1} \sigma_i \sigma_{i+1} \tag{79}$$
   $$\sigma_i \sigma_i^{-1} \leftrightarrow \sigma_i^{-1} \sigma_i \tag{80}$$
   $$\sigma_i \sigma_i^{-1} \sigma_j \leftrightarrow \sigma_j \sigma_i \sigma_i^{-1} \tag{81}$$

   b. If we may, we apply it and store the resultant braid word β in $D_i(A)$ if and only if β is not already contained in $D_j(A)$ for $0 \leq j \leq i$.

   c. We continue to read across α until we have considered all braid words which may be reached from α by a single application of the moves in equations (78) to (81).

   d. Apply steps (a) through (c) for every braid in $D_{i-1}(A)$. If $D_i(A) = \emptyset$, then the algorithm is done.

3. The diagram $D(A)$ of A is the union of all the $D_i(A)$,

$$D(A) = D_0(A) \bigcup D_1(A) \bigcup \cdots \bigcup D_m(A) \tag{82}$$

We show the correctness and termination of this algorithm.

Lemma 4.18. *Algorithm 4.17 terminates for every A and succeeds in listing all braid words B which may be obtained from A by rearranging of generators only, that is using the braid group relations without introducing or removing any generators.*

Proof. $D_0(A)$ is, by definition, finite. It is obvious that for any braid word of finite length, the moves in equations (78) to (81) may be applied a finite number of times. Thus, by induction, every $D_i(A)$ is finite. The number of distinct braid words of a given finite length is finite and since the $D_i(A)$ are, by construction, non-overlapping, their union must be finite. Thus there exists an m such that $D_{m+k}(A) = \emptyset$ for every $k > 0$. Thus the algorithm terminates for every A.

The moves listed in equations (78) to (81) exhaust all possibilities allowed in the braid group under the stipulation that no generators must be removed from or introduced into the word. Thus each word which may be reached from A by rearrangement of generators will eventually be reached by algorithm 4.17 and so the algorithm succeeds in listing all the required braid words. $\qquad\square$

Theorem 4.16 gives the following corollary.

Corollary 4.19. $D(A_{max})$ *contains a braid of the form* EA_m *for* $E \approx e$, *the identity in* B_n.

Proof. By construction $D(A_{max})$ contains all braid words equivalent to A_{max} by rearranging only. By lemma 4.16, A_m can be obtained by a sequence of operations which keeps the length constant or decreases it. Each operation which decreases the length does so by eliminating a sub-word like $e_i = \sigma_i \sigma_i^{-1} \approx \sigma_i^{-1} \sigma_i$.

Since for all i $e_i \approx e$, the identity in B_n, we have

$$e_i \sigma_j^{\pm 1} \approx \sigma_j^{\pm 1} e_i, \qquad e_i e_j \approx e_j e_i \tag{83}$$

for any i and j.

Let us now agree to construct the aforementioned sequence of words without eliminating the sub-words e_i but using equation (83) to bring them all to the left of the word. At the end, we will obtain a word of the form $A^* = EA_m$ where $E \approx e$ is a braid consisting of all these sub-words e_i. The most general form of E is

$$E = e_1^{q_1} e_2^{q_2} \cdots e_{n-1}^{q_{n-1}} \tag{84}$$

with $q_i \geq 0$ for all i. So if we could extract E from A_{max}, we would, in the process, obtain A_m. Since the form EA_m is obtained by rearrangements only, $L(E) \leq L(A^*) = L(A_{max})$. This indicates that $\sum_{i=1}^{n-1} 2q_i \leq L(A_{max})$. $\qquad\square$

Given a braid A, we thus find A_m by constructing the diagram $D(A_{max})$ and selecting the word with the largest number of cancellation pairs such as $\sigma_i \sigma_i^{-1}$. Clearly there will be more than one braid word for the same number of cancellation pairs. We may agree to choose the least braid word lexicographically for definiteness. It is obvious from the construction that this will be a *unique* form of minimal length for the braid A. We thus have an algorithm to find A_m for any A. It is regrettable that the diagram $D(A_{max})$ is, by construction, very large. Two questions are left to ask: Can we make the result stronger and how large is a typical diagram?

In theorem 4.16 we achieved an upper bound for the necessary increase in length of a braid before it may be reduced to a minimum length. One would like to simplify the result somewhat but we shall show in this section that the two straightforward attempts to simplify or strengthen theorem 4.16 are doomed to failure. First we show that we may not, in general, shorten A_{max} to the Garside normal form.

Lemma 4.20. *It is not, in general, possible to obtain A_m from $G(A)$, the Garside normal form of A, by operations which monotonically decrease or keep constant the length of the braid.*

One may think that it would be sufficient to list the diagram of the negative and positive sub-braids of A_{max} and search for a maximal length subbraid which is common to the end of the first and the beginning of the second diagram but this is not true as the following lemma shows.

Lemma 4.21. *There does not exist an A_m in the form $A_1 A_2$ with A_1 negative and A_2 positive for every A.*

Let a be a n-braid of length L with diagram $D(a)$. Consider the braid $a' = a\sigma_i\sigma_i^{-1}$ for some $1 \leq i < n$. We are concerned with the size of $D(a')$ in terms of the size of $D(a)$. For each member of $D(a)$, the cancellation pair $\sigma_i\sigma_i^{-1}$ may appear in any place in both possible orders ($\sigma_i\sigma_i^{-1}$ and $\sigma_i^{-1}\sigma_i$), so in $2(L+1)$ positions. There may be further moves possible by use of the braid group relations but the number of these are clearly bounded by a function linear in L. So the diagram of a word will increase in size by a factor linear in its length for each possible cancellation pair. Given a random positive n-braid a of length L, how many members will $D(a)$ have, on average? We conjecture that:

Conjecture 4.22. For any braid $a \in B_n$ of length L, we have that $|D(a)| \leq |D(\Delta_n^p)|$ with $p = \lceil 2L/(n(n-1)) \rceil$.

Conjecture 4.22 would provide an upper bound for the size of the diagram of any word in terms of the diagrams of the diagrams of Δ_n^p which topologically are a series of p half-twists of the braid strings about the vertical axis. In extensive computer simulations, the conjecture was checked and seems to hold. What it seems to indicate is that the half-twist has the most topological freedom for its length and number of strings under the constraint that the crossing number must be kept constant. This is quite intuitive, yet the conjecture seems to be difficult to prove.

We have investigated the diagrams of several Δ_n^p for their size and for the distribution of braids over the sub-diagrams at each stage of the construction in algorithm 4.17. In table 1 we list the size of the diagram and maximal sub-diagram index for p half-twists on n strings.

Table 1 The Size of Diagrams of Fundamental Words

n	p	$|D(\Delta_n^p)|$	max. i
3	1	2	1
3	2	8	2
3	3	38	5
3	4	196	8
3	5	1062	13
3	6	5948	18
3	7	34120	25
4	1	16	7
4	2	1654	15
5	1	768	25

We conclude that the diagram of a typical braid word grows exponentially with its length and braid index and thus our method of finding the minimal length braid word equivalent to a given braid has exponential complexity. This is not surprising as the problem is NP-Complete. We shall give a heuristic algorithm and other methods later. The properties of the braid groups that made the above solution possible are: (i) It is possible to write all inverse generators as products of the generator of the center and a positive word, (ii) the defining relations relate positive words only and (iii) the braid groups are right and left-cancellative. It is likely that any group which has these properties, has an analog of the Garside normal form and has a solution to the minimum word problem similar to the one above.

Solving the problem exactly is an expensive endeavor and so we ask for approximate methods. It turns out that magnetic relaxation is an important application of this problem and gives rise to good methods to solve it. It is possible to solve the problem heuristically using a purely algebraic algorithm which we now present.

Recall that the braid group B_n is defined by

$$B_n = \langle \ \{\sigma_i\} : 1 \le i < n; \tag{85}$$
$$\sigma_i\sigma_j = \sigma_j\sigma_i \ |i - j| > 1; \sigma_i\sigma_{i+1}\sigma_i = \sigma_{i+1}\sigma_i\sigma_{i+1} \rangle. \tag{86}$$

An n-braid A of c crossings is a word in B_n of word-length c, so the general form of A is

$$A = \sigma_{a_1}^{\epsilon_1}\sigma_{a_2}^{\epsilon_2}\cdots\sigma_{a_c}^{\epsilon_c} \qquad \epsilon_k = \pm 1, \ 1 \le a_k < n, \ \forall k : 1 \le k \le c. \tag{87}$$

Consider an n-braid A of the form given in equation 87. Suppose we wish to find the n-braid A_m equivalent to A such that the length $L(A_m)$ of A_m is minimal over the equivalence class of A. It has been shown [63] that this question is NP-complete and hence computationally difficult (if $P \ne NP$, it is intractable). The following presents a heuristic algorithm for getting close to A_m. We begin with the leftmost generator of A and attempt to move it to the right using both braid group operations. If we can cancel it along the way, we do and if we can not, we move it back to where it started. In this way, we proceed to move all the generators as far to the right as possible. Then we begin at the end and move each generator as far to the left as possible in the same manner. This algorithm will always produce an equivalent braid A' such that $L(A') \leq L(A)$. We consider $L(A)$ generators and move them $O(L(A))$ moves to the right and left. Thus this algorithm takes $O\left(L(A)^2\right)$ time and constant memory. In fact we move a particular generator at most $L(A)$ generators and this is only for the case when all the other generators commute with it, thus the average case complexity is likely to be close to linear in $L(A)$.

This algorithm will not produce a minimum length representative in all cases because it can not unravel complex crossings. To get to the minimum length would require more subtle transformations than just movements to the right or left, which topologically correspond to pulling the strings apart from underneath the crossing. However, as computer experiments show, it does do quite well.

Let us calculate an upper bound to the reduction ratio obtained by this method as a function of n and c. To calculate these, consider the likelihood that a particular generator will be followed by its inverse, which is just $Q_0 = 1/2(n-1)$. The probability Q_j that a generator and its inverse are separated by j generators through which either can be moved is the corresponding probability for $j = 1$ to the power j. We require the number of braids of length 1 which may be generated so as not to contain the generator interfering with the movement of generator σ_i. If $i = 1$ or $n-1$, this is $2(n-3)$ and $2(n-4)$ otherwise. Thus

$$Q_j = \left[\frac{2(n-4)\left(\frac{2(n-1)-2}{2(n-1)}\right) + 2(n-3)\left(\frac{2}{2(n-1)}\right)}{2(n-1)}\right]^j Q_0 \qquad (88)$$

$$= \left[\frac{n^2 - 5n + 5}{(n-1)^2}\right]^j Q_0 \qquad (89)$$

The final factor of Q_0 is present because the generator after the sequence of j generators is required to be inverse of the original generator, an event with probability Q_0. To get the total probability Q of being able to cancel a generator σ_i with its inverse by simple exchange movements over the length $j = 0, 1, \cdots$, we must sum these probabilities in order weighted by the probability that their predecessors did not happen. Thus

$$Q = Q_0 + (1 - Q_0)Q_1 + \cdots + \prod_{k=0}^{j-1}(1 - Q_k)Q_j + \ldots \qquad (90)$$

Note that since the exchange move is not allowed for $n = 3$, $Q = Q_0$ for $n = 3$. The reduction ratio R which occurs as a consequence of this probability is

$R = 1 - 2Q$ since each time that the event happens two generators may be canceled. Note that in this calculation we have considered the probability that a generator can be moved next to its inverse in the word using only the far commutation relation that $\sigma_i \sigma_j = \sigma_j \sigma_i$ for $|i - j| > 1$ in a long braid. The heuristic algorithm however uses both braid group moves to attempt to move generators next to their inverses. Thus R is an upper bound for the reduction ratio achieved by the heuristic algorithm as the braid becomes long.

In §6 we present the results of the algebraic reduction of a large number of braids but a few comments about the efficiency of the algorithm are in order. The only exact algorithm to minimize braid is valid only for $n \leq 3$ [11] and by comparing this heuristic to this exact algorithm, we find that the heuristic finds a braid the length of which is within five percent of the length found by the exact algorithm and that it reaches the actual minimum in 0.005 of all cases. This shows that the heuristic is quite effective for $n = 3$ (note that reduction for $n = 1, 2$ is trivial since B_1, B_2 are free groups).

5 Open Problems

The problems that follow are, to the best of my knowledge, unsolved at the time of publication. The selection is personal and far from complete. However, each problem is quite significant in that its solution will have an impact on research. I believe that even partial results for most of these problems would be worth a Ph.D. The problems are presented in no particular order.

1. Is the Burau representation faithful for $n = 4$? Seems hard to answer but the significance is somewhat debatable since it is unfaithful for $n > 4$.
2. Does the Jones polynomial identify the unknot? In other words, is there a knot (distinct from the unknot) that has the same Jones polynomial as the unknot? (Warning: Searches for a counterexample have gotten quite far in the knot tables, so if there is a counterexample it is not simple. At present, it seems likely that there is no counterexample.)
3. Construct a polynomial-time algorithm for calculating the Jones and related polynomials or show that this cannot be done (assuming that $P \neq NP$ a proof of NP-completeness would do).
4. Construct an inherently algebraic algorithm for solving the Markov problem. (Note: The problem is definitely solvable via Hemions algorithm but this is very complicated, we are looking for a simple algorithm even if it is in exponential-time.)
5. Prove that the Markov problem is NP-complete, intractable or find a polynomial-time algorithm for it. (Note: Depending on how this is done and the answer, this is probably a Fields Medal problem but that should not discourage you from trying it.)

6. Construct a provably secure cryptosystem on the basis of the braid or similar groups. (Note: With present systems the question of security is not quite settled.)

7. Create a general theory of 2-tangle equations and how to solve them. (Note: The solvability needs to be practical as these equations actually come up in biology.) As an encore, do it for n-tangle equations.

8. Create a theory of physical braids with aims to study then tension properties of various knots independent of the rope on which they are tied.

9. Construct further invariants of knots and braids, preferably easily calculable and strong (not too many distinct knots having the same value of your invariant).

10. Construct a complete invariant of knots that can actually be compared computationally. (Note: The peripheral group system is complete but it cannot be compared to other systems because groups cannot in general be distinguished from another.)

11. Build a software system that makes it easy to input/output knots and braids, compute many invariants, identify the knot in a table, construct a knot table and play with new ideas largely independent of complex programming.

References

1. Adams, C., Hildebrand, M., Weeks, J. (1991): Hyperbolic Invariants of Knots and Links. Trans. Amer. Math. Soc., **326**, 1–56.
2. Adian, S. I. (1957): The unsolvability of certain algorithmic problems in the theory of groups. Trudy Moskov. Mat. Obsc., **6**, 231–298.
3. Adian, S. I. (1957): Finitely Presented Group and Algorithms. Dokl. Akad. Nauk SSSR, **117**, 9–12.
4. Alexander, J. W. (1923): A lemma on systems of knotted curves. Proc. Nat. Acad. Sci. USA, **9**, 93–95.
5. Appel, K. I. and Schupp, P. E. (1972): The Conjugacy Problem for the Group of any Tame Alternating Knot is Solvable. Proc. Amer. Math. Soc., **33**, 329–336.
6. Artin, E. (1925): Theorie der Zöpfe. (in German) Abh. Math. Sem. Univ. Hamburg, **4**, 47–72.
7. Artin, E. (1947): Theory of braids. Ann. Math., **48**, 101–126.
8. Baader, F., Nipkow, T. (1998): Term Rewriting and All That. (Cambridge University Press, Cambridge).
9. Bangert, P. D., Berger, M. A., Prandi, R. (2002): In Search of Minimal Random Braid Configurations. J. Phys. A., **35**, 43–59.
10. Berger, M. A. (1993): Energy-crossing number relations for braided magnetic fields. Phys. Rev. Lett., **70**, 705–708.
11. Berger, M. A. (1994): Minimum crossing numbers for 3-braids. J. Phys. A, **27**, 6205–6213.
12. Berger, M.A. (2000): Hamiltonian dynamics generated by Vassiliev invariants. J. Phys. A., **34**, 1363–1374.
13. Bigelow, S. (1999): The Burau representation is not faithful for $n = 5$. Geometry and Topology, **3**, 397–404.

14. Birman, J. S. (1974): Braids, Links and Mapping Class Groups. Ann. of Math. Studies 82 (Princeton Univ. Press, Princeton).

15. Birman, J. S., Ko, K. H., Lee, S. J. (1998): A New Approach to the Word and Conjugacy Problems in the Braid Groups. Ad. Math., **139**, 322–353.

16. Birman, J. S., Menasco, W. (1992): Studying Links Via Closed Braids I: A Finiteness Theorem. Pacific J. Math., **154**, 17–36.

17. Birman, J. S., Menasco, W. (1991): Studying Links Via Closed Braids II: On a Theorem of Bennequin. Topology and Its Applications, **40**, 71–82.

18. Birman, J. S., Menasco, W. (1993): Studying Links Via Closed Braids III: Classifying Links which are Closed 3-braids. Pacific J. Math., **161**, 25–113.

19. Birman, J. S., Menasco, W. (1990): Studying Links Via Closed Braids IV: Closed Braid Representatives of Split and Composite Links. Invent. Math., **102**, 115–139.

20. Birman, J. S., Menasco, W. (1992): Studying Links Via Closed Braids V: Closed Braid Representatives of the Unlink. Trans. AMS, **329**, 585–606.

21. Birman, J. S., Menasco, W. (1992): Studying Links Via Closed Braids VI: A Non-Finiteness Theorem. Pacific J. Math., **156**, 265–285.

22. Birman, J. S., Menasco, W. (1992): A calculus on links in the 3-sphere. in Kawauchi, A. *Knots 90* (Walter de Gruyter, Berlin), 625–631.

23. Birman, J. S., Wajnryb, B. (1986): Markov Classes in Certain Finite Quotients of Artin's Braid Group. Israel J. Math., **56**, 160–178.

24. Birkhoff, G. (1935): On the structure of abstract algebras. Proc. Camb. Phil. Soc., **31**, 433–454.

25. Bohnenblust, F. (1947): The Algebraical Braid Group. Ann. Math., **46**, 127–136

26. Boyland, P. L., Aref, H., Stremler, M. A. (2000): Topological fluid mechanics of stirring. J. Fluid Mech., **403**, 277–304.

27. Buchberger, B. (1987): History and Basic Features of the Critical-Pair/Completion Procedure. J. Sym. Comp. **3**, 3–38. Printed in *Rewriting Techniques and Applications* ed. by Jouannaud, J.-P. (Academic Press, London), 3–38.

28. Chan, T. (2000): HOMFLY polynomials of some generalized Hopf links. J. Knot Th. Rami., **9**, 865–883.

29. Cohen, D.E. (1987): Computability and Logic. (Ellis Horwood, Chichester).

30. Conway, J.H. (1970): An Enumeration of Knot and Links, and Some of their Algebraic Properties. in Leech, J. (1970): Computational Problems in Abstract Algebra. (Pergamon, Oxford)

31. Coxeter, H.S.M., Moser, W.O.J. (1957): Generators and Relations for Discrete Groups. (Springer, Berlin)

32. Dershowitz, N. (1979): A note on simplification orderings. Inform. Proc. Let., **9**, 212–215.

33. Dershowitz, N. (1981): Termination of linear rewriting systems. in *Automata, Languages and Programming* ed. by Even, S. and Kariv, O., Lecture Notes in Computer Science Volume 115 (Springer, Heidelberg), 448–458.

34. Dershowitz, N. (1987): Termination of Rewriting. J. Sym. Comp., **3**, 69–116. Printed in *Rewriting Techniques and Applications* ed. by Jouannaud, J.-P. (Academic Press, London), 69–116.

35. Dowker, C. H. and Thistlethwaite, M. B. (1983): Classification of Knot Projections. Top. Appl., **16**, 19–31.

36. Epstein, D.B.A., Cannon, J.W., Holt, D.F., Levy, S.V.F., Paterson, M.S., Thurston, W.P. (1992): Word Processing in Groups. (Jones and Bartlett, Boston).

37. Garside, F.A. (1969): The braid group and other groups. Quart. J. Math. Oxford, **20**, 235–254.

38. Gilbert, N.D., Porter, T. (1994): Knots and Surfaces. (Oxford University Press, Oxford).

39. Hemion, G. (1992): The Classification of Knots and 3-dimensional Spaces. (Oxford Univ. Press, Oxford).

40. Hempel, J. (1976): 3-manifolds. Ann. of Math. Studies Volume 86 (Princeton Uni. Press, Princeton).

41. Huet, G. (1980): Confluent reductions: Abstract properties and applications to term rewriting systems. J. Assoc. Comput. Mach., **27**, 797–821.

42. Huet, G. (1981): A Complete Proof of the Knuth-Bendix Completion Algorithm. J. Comp. Syst. Sci., **23**, 11–21.

43. Huet, G., Lankford, D.S. (1978): On the uniform halting problem for term rewriting systems. Rapport laboria 283, Institut de Recherche en Informatique et en Automatique, Le Chesnay, France.

44. Jacquemard, A. (1990): About the effective classification of conjugacy classes of braids. J. Pure Appl. Al., **63**, 161–169.

45. Johnson, D.L. (1980): Topics in the Theory of Group Presentations. Lon. Math. Soc. Lec. Notes Vol. 42 (Cambridge Uni. Press, Cambridge).

46. Jouannoud, J.P., Kirchner, H. (1986): Completion of a set of rules modulo a set of equations, SIAM. J. Comp., **15**, 1155–1194.

47. Kang, E. S. *et. al.* (1997): Band-generator presentation for the 4-braid group. Top. Appl., **78**, 39–60.

48. Kauffman, L. (1993): Knots and Physics. Series on Knots and Everything Vol. 1. World Scientific, Singapore.

49. Kawauchi, A. (1996): A Survey of Knot Theory. (Birkhäuser Verlag, Basel)

50. Knuth, D.E., Bendix, P.B. (1970): Simple word problems in universal algebras. in *Computational Problems in Abstract Algebra* ed. by Leech, J. (Pergamon Press, Oxford), 263–297. Reprinted in 1983 in *Automation of Reasoning 2* (Springer, Berlin), 342–376.

51. Lambropoulou, S.S.F. (1993): A Study of Braid in 3-manifolds. unpublished PhD thesis (Univ. of Warwick).

52. Lambropoulou, S.S.F., Rourke, C.P. (1997): Markov's Theorem in 3-manifolds. Top. Appl., **78**, 95–22.

53. The CiME system is available from http://cime.lri.fr.

54. Magnus, W., Peluso, A. (1967): On Knot Groups. Comm. Pure Appl. Math., **20**, 749–770.

55. Markov, A.A. (1935): Über die freie Äquivalenz geschlossener Zöpfe (in German), Recueil Mathematique Moscou, **1**, 73–78. [Mat. Sb. **43** 1936.]

56. McCool, J. (1980): On Reducible Braids. in *Word Problems II* ed. by Adian, S. I., Boone, W. W. and Higman, G. (North-Holland, Amsterdam), 261–295.

57. Miller, C.F. III (1992): Decision Problems for Groups - Survey and Reflections. in *Algorithms and Classification in Combinatorial Group Theory* ed. by Baumslag, G. and Miller, C. F. III, Math. Sci. Re. Inst. Pub. Volume 23 (Springer, New York), 1–59.

58. Morton, H.R. (1983): An irreducible 4-string braid with unknotted closure. Math. Proc. Camb. Phil. Soc., **93**, 259–261.

59. Morton, H.R. (1986): Threading Knot Diagrams. Math. Proc. Camb. Phil. Soc., **99**, 247–260.

60. Murasugi, K. (1996): Knot Theory And Its Applications. (Birkhäuser, Boston).

61. Murasugi, K., Thomas, R.S.D. (1972): Isotopic closed nonconjugate braids. Proc. Am. Math. Soc., **33**, 137–139.

62. Newman, M.H.A. (1942): On theories with a combinatorial definition of 'equivalence'. Ann. Math., **43**, 223–243.

63. Paterson, M.S., Razborov, A.A. (1991): The set of minimal braids is co-NP-complete. J. Algorithms, **12**, 393–408.

64. Rabin, M.O. (1958): Recursive Unsolvability of Group Theoretic Problems. Ann. Math., **67**, 172–194.

65. Reidemeister, K. (1983): Knot Theory. BCS Associates, Moscow, Idaho. Originally published as Reidemeister, K. (1932): Knotentheorie. Springer, Berlin

66. Ricca, R.L. (1998): Applications of Knot Theory in Fluid Mechanics. in Jones, V.F.R. *et. al.*, ed. by *Knot Theory* Banach Center Pub. Vol. 42 (Inst. of Math., Polish Acad. Sci., Warszawa), 321–346.

67. Schubert, H. (1948): Die Eindeutige Zerlegbarkeit eines Knotens in Primknoten. (in German), Sitz. Heidelberger Akad. Wiss., math.-nat. Kl., 55–104.
68. Schubert, H. (1961): Bestimmung der Primfaktorzerlegung von Verkettungen. (in German), Math. Zeitschr., **76**, 116–148.
69. Tatsuoka, K. (1987): Geodesics in the braid group. preprint, Dept. of Mathematics, University of Texas at Austin.
70. Tourlakis, G.J. (1984): Computability. (Reston, Reston).
71. Turing, A.M. (1937): On Computable Numbers, with an Application to the Entscheidungsproblem. Proc. London Math. Soc. Ser. 2, **42**, 230–265.
72. Vogel, P. (1990): Representation of links by braids: A new algorithm. Comment. Math. Helvetici, **65**, 104–113.
73. Waldhausen, F. (1968): On Irreducible 3-manifolds which are Sufficiently Large. Ann. Math., **87**, 56–88.
74. Waldhausen, F. (1968): The Word Problem in Fundamental Groups of Sufficiently Large Irreducible 3-manifolds. Ann. Math., **88**, 272–280.
75. Williams, R.F. (1988): The Braid Index of an Algebraic Link. In: Birman, J. S., Libgober, A. (ed) Braids. Amer. Math. Soc., Providence.
76. Xu, P.J. (1992): The genus of closed 3-braids. J. Knot Theory Ramifications, **1**, 303–326.
77. Yamada, S. (1987): The minimal number of Seifert circles equals the braid index of a link. Invent. Math., **89**, 347–356.
78. Yoder, M.A. (1995): String Rewriting Applied to Problems in the Braid Groups. unpublished Ph.D. thesis (Uni. South Florida).

Topological Quantities: Calculating Winding, Writhing, Linking, and Higher order Invariants

Mitchell A. Berger
(CIME Lecturer)

Abstract Many topological calculations can be done most easily using the basic idea of *winding number*. This chapter demonstrates the use of winding number techniques in calculating writhe, linking number, twist, and higher order braid invariants. The writhe calculation works for both closed and open curves. These measures have applications in molecular biology, materials science, fluid mechanics and astrophysics.

1 Introduction

Often we can find several equivalent expressions for a mathematical quantity. One of these may seem most fundamental, and so may be chosen to be the definition of the quantity. Other expressions can prove useful in proving theorems or in calculations. This chapter considers geometrical quantities associated with curves, such as twist, writhe, linking numbers and higher order invariants.

For all these quantities we can obtain expressions based on the fundamental idea of *winding number*. This chapter will show how winding number techniques simplify and speed up calculations, and add insight into what the quantities actually measure.

Objects such as knots and links are fundamentally three-dimensional – and yet many knot invariants rely on almost two-dimensional representations. The usual method involves projecting knotted curves onto a plane. Information about the third dimension only survives at crossings: at places where one

M.A. Berger
Mathematics Dept., SECAM, University of Exeter,
North Park Rd. Exeter EX4 4QE, U.K.
e-mail: m.berger@exeter.ac.uk
http://secamlocal.ex.ac.uk/people/staff/mab215/home.htm

R.L. Ricca (ed.), *Lectures on Topological Fluid Mechanics*,
Lecture Notes in Mathematics 1973, DOI: 10.1007/978-3-642-00837-5,
© Springer-Verlag Berlin Heidelberg 2009

part of a curve crosses under or over another part we specify which curve is on top. Invariants can then be deduced by examining the pattern of over and under-crossings.

This chapter investigates a less severe reduction of dimensionality. Here one direction (which we will generally take to be the z direction) is singled out. Suppose s measures arclength along a curve. Then a three-dimensional curve can be parameterized by s, and specified by the three coordinate functions $(x(s), y(s), z(s))$. If, however, $z(s)$ is a monotonic function of s, i.e. the curve always travels upwards (or downwards) in z, then we could parameterize by z instead. Now we need only two coordinate functions – $(x(z), y(z))$. This considerably simplifies calculations. Also, this procedure immediately makes a connection with two–dimensional dynamics: simply replace the letter z with the letter t, symbolizing time. The 3–D curve $(x(z), y(z), z)$ becomes the time history (space–time diagram) $(x(t), y(t), t)$ of a point moving in the $x - y$ plane.

The reader may already be objecting that most curves (certainly all closed curves) are not monotonic in z: they go up and down. To deal with this simple fact, we will have to chop up our curves at turning points where $dz/ds = 0$ (see figure 1). This is the price we must pay for the reduction from three coordinate functions to two. We will now have to sum over all the pieces of our curves when we wish to calculate something like writhe or linking number.

In many applications a special direction is already singled out. For example, elastic rod experiments sometimes fix the rod between parallel planes. In some astrophysical applications, one considers the geometry and topology of magnetic field lines anchored at a boundary surface. This surface might be

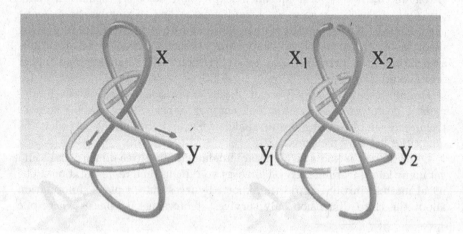

Fig. 1 On the left, a Whitehead link consisting of two curves $\mathbf{x}(t)$ and $\mathbf{y}(t)$. Arrows give the directions of the curves, i.e. the directions of increase of arclength s. On the right, the curves have been chopped into sections $\mathbf{x}_1, \mathbf{x}_2$ and $\mathbf{y}_1, \mathbf{y}_2$ at their maxima and minima in z.

the photosphere of the sun or the surface of an accretion disk. In the solar case z should be replaced by radial coordinate r. In other applications we start with a time history of the two–dimensional motion of several objects (e.g. [KHS06, GVV03]). In the corresponding space–time diagram the curves $(x(t), y(t), t)$ twist and braid about each other; the topology and geometry of this braiding gives information about the dynamics.

2 Winding Numbers

2.1 Two Braided Curves between Parallel Planes

First consider two curves stretching between parallel planes (see figure 2) separated by a height h. Let these curves be labelled $\mathbf{x}(z)$ and $\mathbf{y}(z)$. Also let $\mathbf{r}(z) = \mathbf{y}(z) - \mathbf{x}(z)$ be the relative position vector pointing from $\mathbf{x}$ to $\mathbf{y}$. This vector is horizontal; we will be interested in how much it rotates as it rises from the bottom to the top. Recall from elementary mechanics that the angular velocity $d\Theta/dt$ of a particle moving in the $x - y$ plane with position

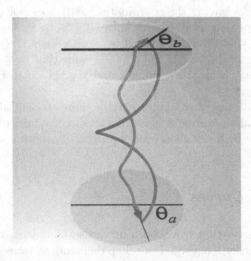

Fig. 2 Two braided curves $\mathbf{x}$ and $\mathbf{y}$. Both curves have endpoints on horizontal planes $z = a$ and $z = b$. We assume both curves are oriented upwards, i.e. $dz/ds > 0$ where s denotes arclength. The horizontal axis is shown on both planes. Also, a relative position vector has been drawn on each plane between the endpoints of curve $\mathbf{x}$ and curve $\mathbf{y}$. The angles between these vectors and the x axis are labelled Θ_a and Θ_b. For these curves, the net winding number is $w = 1 + (\Theta_b - \Theta_a)/2\pi$. The extra unit arises because the branch cut in angle has been crossed, in this case because the curves wrap by more than 2π.

vector $\mathbf{r}(t)$ is given by

$$\frac{d\Theta}{dt} = \frac{(\mathbf{r} \times d\mathbf{r}/dt)_z}{r^2}. \tag{1}$$

Similarly the rotation rate of our two curves as they go upwards is

$$\frac{d\Theta}{dz} = \frac{(\mathbf{r} \times d\mathbf{r}/dz)_z}{r^2}. \tag{2}$$

Let $\Theta(z = a) = \Theta_a$ be the orientation of $\mathbf{r}(a)$ with respect to the x axis, i.e. $\mathbf{r}(a) = r_a(\cos\Theta_a, \sin\Theta_a)$. We place the discontinuity (branch cut) in Θ on the negative x axis, i.e. $-\pi < \Theta \le \pi$. For the winding number, we count a complete turn as one unit, so the net winding is given by

$$w_{\mathbf{xy}} = \frac{1}{2\pi}\left(\int_a^z \frac{d\Theta}{dz'}\,dz'\right); \tag{3}$$

$$= \frac{1}{2\pi}(\Theta_b - \Theta_a) + n, \tag{4}$$

where n is an integer counting the number of times the angle rotates through the branch cut in the anti-clockwise direction.

Note that $w_{\mathbf{xy}}$ is independent of choice of x axis: a rotation of coordinates through $\delta\theta$ in the $x \to y$ plane (or alternatively a uniform rotation of the braid through $-\delta\theta$ leaving the coordinates fixed) does not change $w_{\mathbf{xy}}$. To see this, first note that away from branch cuts, both Θ_b and Θ_a change by $\delta\theta$, so equation (4) stays fixed. Also, if (say) Θ_b crosses the cut at $\pm\pi$, then it will jump by $\pm2\pi$, but n will jump at the same time by ∓1.

In addition, $w_{\mathbf{xy}} = w_{\mathbf{yx}}$: reversing $\mathbf{x}$ and $\mathbf{y}$ adds $\pm\pi$ to both Θ_b and Θ_a without changing n.

Finally, we note that a computer may be more efficient at simply detecting crossings of the negative x axis than in directly integrating $d\Theta/dz$, so equation (4) may be faster than equation (3) in computing w.

2.2 General Curves

Next consider curves that travel both upwards and downwards; for example the two closed curves in figure 1. If we chop them into sections at their maxima and minima then we can ask about the winding number between each pair of sections. Figure 3 shows the two sections $\mathbf{x}_2$ and $\mathbf{y}_2$. These sections overlap between $z_{\min}$ and $z_{\max}$. Both sections travel downwards; i.e. $dz/ds < 0$ for both. Let σ denote the sign of dz/ds, and let

$$w_{22} = \frac{\sigma(\mathbf{x}_2)\sigma(\mathbf{y}_2)}{2\pi}\int_{z_{\min}}^{z_{\max}} \frac{d\Theta}{dz}\,dz. \tag{5}$$

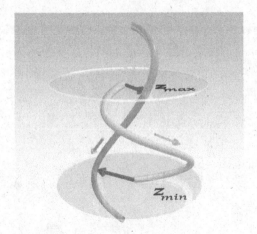

Fig. 3 The two sections $x2$ and $y2$ of the curves in figure 1. The relative position vectors are shown at the maximum and minimum heights z_{max}, z_{min} where these two sections overlap.

Then for these curves

$$w_{22} = \frac{1}{2\pi}(\Theta_{max} - \Theta_{min}) \tag{6}$$

(for this diagram $\Theta(z)$ does not cross a branch cut, and so $n = 0$).

In the example shown in figure 4, curve 1 is monotonic, but curve 2 has a maximum at z_2 and a minimum at z_3, so must be chopped into three pieces, $j = 1, 2, 3$. Let w_{ij} measure the net winding of piece i of curve 1 with piece j of curve 2. The first curve has just one piece, so we can write

$$w = w_{11} + w_{12} + w_{13} \tag{7}$$

where

$$w_{11} = \frac{1}{2\pi}\int_a^{z_2} \frac{d\Theta}{dz}\,dz; \qquad w_{12} = -\frac{1}{2\pi}\int_{z_3}^{z_2} \frac{d\Theta}{dz}\,dz; \tag{8}$$

$$w_{13} = \frac{1}{2\pi}\int_{z_3}^b \frac{d\Theta}{dz}\,dz. \tag{9}$$

The integrals give

$$w_{11} = \frac{1}{2\pi}(\Theta(z_2) - \Theta_a) + n_{11}; \tag{10}$$

$$w_{12} = -\frac{1}{2\pi}(\Theta(z_2) - \Theta(z_3)) + n_{12}; \tag{11}$$

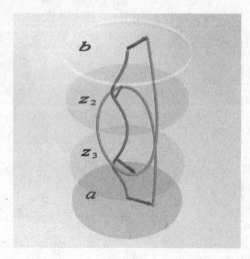

Fig. 4 Two tangled curves. The curve on the right starts at $z = a$, rises to a local maximum at $z = z_2$, wraps around the second curve as it descends to a local minimum at $z = z_3$, then rises to $z = b$. We will chop this curve at heights z_2 and z_3. The relative position vectors $\mathbf{r}(z)$ are shown at $z_1 = a$, z_2, z_3, and $z_4 = b$. For these curves the total winding number is $w = 1 + (\Theta_b - \Theta_a)/2\pi$. Note that w does not depend on the intermediate angles $\Theta(z_2)$ or $\Theta(z_3)$.

$$w_{13} = \frac{1}{2\pi}(\Theta_b - \Theta(z_3)) + n_{13}; \tag{12}$$

$$w = \frac{1}{2\pi}(\Theta_b - \Theta_a) + n_{11} + n_{12} + n_{13}. \tag{13}$$

Note that the intermediate angles $\Theta(z_2)$ and $\Theta(z_3)$ cancel out in the total winding number w.

We can now construct general expressions for the winding numbers between two arbitrary curves. For each section of a curve we define a function $\sigma(z)$ which first tells us whether the piece exists at z, and if so, then tells us whether it is rising or falling, i.e. gives the sign of dz/ds. In general, we will only consider curves with a finite set of discrete points where $dz/ds = 0$. Say curve α has n of pieces in between these turning points. Piece i starts at some height z_i and ends at height z_{i+1}. Define

$$\sigma_{\alpha i}(z) = \begin{cases} 1, & z \in (z_i, z_{i+1}) \text{ and } dz/ds0; \\ -1, & z \in (z_i, z_{i+1}) \text{ and } dz/ds < 0; \\ 0, & z \notin (z_i, z_{i+1}). \end{cases} \tag{14}$$

For example, the curves in figure 4 have winding number

$$w = \sum_{j=1}^{3} w_{1j}; \qquad w_{1j} = \sigma_1 \sigma_{2j} \frac{1}{2\pi} \int_a^b \frac{d\Theta}{dz} \, dz. \tag{15}$$

Let us define a generalized winding number $w[a, b]$ measuring the rotation of two curves between heights $z = a$ and $z = b$: if the curves can be cut into m_1 and m_2 pieces, then

$$w[a, b] = \sum_{i=1}^{m_1} \sum_{j=1}^{m_2} w_{ij}[a, b]; \qquad w_{ij}[a, b] = \sigma_{1i}\sigma_{2j}\frac{1}{2\pi}\int_a^b \frac{d\Theta_{ij}}{dz}\,dz. \qquad (16)$$

Furthermore, if pieces $1i$ and $2j$ overlap between heights $z_{\min}(1i, 2j)$ and $z_{\max}(1i, 2j)$, and we let $\Theta_{ij\,\min}$ and $\Theta_{ij\,\max}$ be the orientations of the relative position vectors at these heights, then

$$w_{ij}[a, b] = \sigma_{1i}\sigma_{2j}\left(\frac{1}{2\pi}(\Theta_{ij\,\max} - \Theta_{ij\,\min}) + n_{ij}\right). \qquad (17)$$

2.3 Topological Invariance

Consider two curves with fixed endpoints on boundary planes, as in figure 2 and figure 4. Suppose we distort the curves *between* the two planes. As the endpoints are fixed, the boundary angles Θ_a and Θ_b do not change. But w equals $(\Theta_a - \Theta_b)/2\pi$ plus an integer. So w can only change by integer amounts due to the motion of the curves. If the two curves are allowed to pass through each other, then w will indeed change by ± 1 during each pass-through.

We will require that the curves do not pass through the boundary planes except at the endpoints. To see why, suppose one of the curves forms a loop which rises through the top plane. This loop can then jump over the top endpoint of the other curve. If the loop is then brought back downwards the new configuration will have a change in the winding number of ± 1.

If the two curves are not allowed to pass through each other or loop-over, then a discrete change in w is not possible. To see this, note that equation (16) expresses w as a sum of integrals involving $d\Theta/dz$ (see equation equation (2)). Now $d\Theta/dz$ includes $|\mathbf{r}|^{-2}$, which goes singular if the curves pass through each other. Thus it is not surprising that w can jump in this situation. But with material strings, or simply the restriction that $|\mathbf{r}| > 0$ at all times, the integrals evolve continuously, and jumps are not possible. These considerations suggest the following theorem:

Theorem 2.1 (Invariance of Winding Numbers). . *Suppose two curves stretch between the planes $z = a$ and $z = b$, where each curve has endpoints on both planes. Then their winding number $w[a, b]$ is invariant to continuous deformations of the curves in the region $a < z < b$ where*

1. *the endpoints are fixed;*
2. *the curves cannot pass through the boundary plane (except at their endpoints); and*
3. *the curves do not pass through each other.*

3 Linking Numbers

3.1 *Winding Number Derivation*

Next let us calculate the net winding number for two closed curves. Suppose the curves overlap in height between $z_{\min}$ and $z_{\max}$. Suppose we define the quantity

$$\mathcal{L}_k = w[-\infty, \infty] = w[z_{\min}, z_{\max}]. \tag{18}$$

This quantity will turn out to be the *linking* number between the two curves.

Suppose, for example, that each curve has only one maximum and one minimum, as in figure 5. Generically, the two maxima and minima are at different heights. We again consider the orientation angles Θ of the relative position vectors $\mathbf{r}(z)$. The net winding angle will depend on two orientation angles Θ_A, Θ_B at the lower of the two maxima, and two orientation angles Θ_C, Θ_D at the upper of the two minima. We have

$$w_{11} = \frac{1}{2\pi}(\Theta_A - \Theta_C) + n_{11}; \tag{19}$$

$$w_{12} = -\frac{1}{2\pi}(\Theta_B - \Theta_D) + n_{12}; \tag{20}$$

$$w_{21} = -\frac{1}{2\pi}(\Theta_A - \Theta_C) + n_{21}; \tag{21}$$

$$w_{22} = \frac{1}{2\pi}(\Theta_B - \Theta_D) + n_{22}; \tag{22}$$

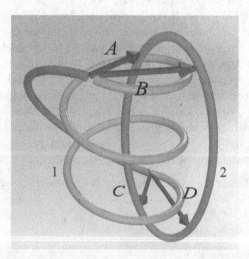

Fig. 5 Two closed curves. The relative position vectors $\mathbf{r}_A, \mathbf{r}_B, \mathbf{r}_C, \mathbf{r}_D$ are shown at the maximum and minimum points on curve 1, $z_{1\,\max}$ and $z_{1\,\min}$.

$$W = \sum_{i,j=1}^{2} w_{ij} = n_{11} + n_{12} + n_{21} + n_{22}. \tag{23}$$

All the principal angles cancel out, leaving us with an integer

$$\mathcal{L}_k = \sum_{i,j} n_{ij}. \tag{24}$$

The linking number is invariant to any distortion of the curves (without letting them pass through each other). First, an integer cannot continuously change to another integer! Only discrete changes are conceivable – for example, if the two curves do pass through each other then that would constitute a discrete event. Now, in the winding number calculation, the integer counts how many times the relative position angles $\Theta_{ij}(z)$ pass through the branch cut. One could rotate parts of the curves to remove any particular branch cut pass – however, another branch pass will just be created elsewhere. This is because the relative position angle must change through a whole range of 2π to increase the winding number by $+1$.

3.2 General Properties

Let the tangent vector to a curve $\mathbf{x}$ (parametrized by arclength s) be defined by

$$\widehat{\mathbf{T}}(s) = \frac{d\mathbf{x}}{ds}. \tag{25}$$

A tangent vector to a curve parameterized by arclength has unit norm, so $|\widehat{\mathbf{T}}(s)| = 1$.

The Gauss linking number was originally defined as a double integral over two closed curves $\mathbf{x}$ (with points $\mathbf{x}(s)$ and tangent $\mathbf{T_x}(s)$) and $\mathbf{y}$ (with points labelled $\mathbf{y}(s')$ and tangent $\mathbf{T_y}(s')$):

$$\mathcal{L}_k \equiv \frac{1}{4\pi} \oint_{\mathbf{x}} \oint_{\mathbf{y}} \mathbf{T_x}(s) \times \mathbf{T_y}(s') \cdot \frac{\mathbf{x}(s) - \mathbf{y}(s')}{|\mathbf{x}(s) - \mathbf{y}(s')|^3} \, ds \, ds'. \tag{26}$$

In addition to being an integer and a topological invariant, linking number possesses the following properties:

1. $\mathcal{L}_k$ equals half the signed number of crossings of the two curves as seen in any plane projection (see figure 5).
2. $\mathcal{L}_k$ changes to $-\mathcal{L}_k$ if one of the curves changes direction ($\widehat{\mathbf{T}} \to -\widehat{\mathbf{T}}$). It thus is invariant to a change in direction of both curves.

4 Twist and Writhe Numbers

4.1 Ribbons

Suppose we compute the linking number of two curves which closely track
each other. We will call one of the curves the *axis curve* **x**; the other will be
the *secondary curve* **y**. For example, the two curves might be the bounding
curves of a ribbon, or the central axis and one of the two strands of a DNA
double helix.

Suppose we parameterize the axis by some variable t (popular choices
might be $t = z$, vertical height, or $t = s$, arclength); then we will let $\widehat{\mathbf{V}}(t)$ be
a unit vector normal to $\widehat{\mathbf{T}}(t)$ pointing towards $\mathbf{y}(t)$. In particular,

$$0 = \widehat{\mathbf{V}}(t) \cdot \widehat{\mathbf{T}}(t); \tag{27}$$

$$\mathbf{y}(t) = \mathbf{x}(t) + \epsilon(t)\widehat{\mathbf{V}}(t). \tag{28}$$

Here we will assume that the distance $\epsilon(t)$ between the two curves is small;
for example smaller than the radius of curvature of $\mathbf{x}(t)$. Note that, while we
can parameterize both curves by $t = s$, the parameter s measures arclength
along the axis curve, not along the secondary curve.

The Gauss linking integral, equation (26), can be computed as usual. But
note the term in the denominator. If we let $\epsilon \to 0$, so that the two curves
coincide, then this denominator can vanish for the set of points $s = s'$. As this
is only a one-dimensional subset of the two dimensional set of points s, s', the
singularity is not fatal. Călugăreanu [C59] studied this limit in detail, and
showed that the linking number (at finite but small ϵ) can be decomposed
into two terms:

$$\mathcal{L}_k = \mathcal{W}_r + \mathcal{T}_w. \tag{29}$$

Here the *writhe* number $\mathcal{W}_r$ is the limiting integral (when both line integrals
go along the axis curve),

$$\mathcal{W}_r \equiv \frac{1}{4\pi} \oint_{\mathbf{x}} \oint_{\mathbf{x}} \widehat{\mathbf{T}}(s) \times \widehat{\mathbf{T}}(s') \cdot \frac{\mathbf{x}(s) - \mathbf{x}(s')}{|\mathbf{x}(s) - \mathbf{x}(s')|^3} \, ds \, ds', \tag{30}$$

while the *twist* number $\mathcal{T}_w$ is defined by

$$\mathcal{T}_w \equiv \frac{1}{2\pi} \oint_{\mathbf{x}} \widehat{\mathbf{T}}(s) \cdot \widehat{\mathbf{V}}(s) \times \frac{d\widehat{\mathbf{V}}(s)}{ds} \, ds \tag{31}$$

$$= \frac{1}{2\pi} \oint_{\mathbf{x}} \frac{1}{|\mathbf{v}(s)|^2} \widehat{\mathbf{T}}(s) \cdot \mathbf{v}(s) \times \frac{d\mathbf{v}(s)}{ds} \, ds \tag{32}$$

where $\mathbf{v} = \epsilon\widehat{\mathbf{V}}$.

Subsequent research has illuminated the meaning and properties of these two new quantities (see e.g. [F78, MR92, AKT95, C05]). As a quick (but not always accurate) guide, the writhe of a curve measures how much it kinks and coils, while twist measures how much a secondary curve twists about the first. The main application of these terms has been to DNA studies [VM97]. A DNA molecule has two strands which wind about each other thousands of times. Some of the winding number (or linking number if the DNA is closed) arises from local twist of the strands about the axis of the molecule, and the remainder arises because of coiling of the axis itself. As DNA can be centimetres long if fully stretched out, it needs a great deal of coiling to fit into a microscopic cell!

Twist has some special properties:

1. First, linking number and writhe are double integrals – while $\mathcal{T}_w$ is a single integral. Thus it is meaningful to write twist as the sum of a local density along the curve:

$$\mathcal{T}_w = \frac{1}{2\pi} \oint_{\mathbf{x}} \frac{d\mathcal{T}_w}{ds}\,ds. \tag{33}$$

2. $d\mathcal{T}_w/ds$ measures the rotation rate of the secondary curve about the axis curve, in the plane perpendicular to the tangent vector $\mathbf{T_x}(s)$. Recall that equation (2) gives the angular velocity of a curve about the constant z direction. If instead we wished to find the angular velocity about the tangent direction, we would calculate

$$\frac{d\mathcal{T}_w}{ds} = \frac{1}{2\pi} \widehat{\mathbf{T}}(s) \cdot \widehat{\mathbf{V}}(s) \times \frac{d\widehat{\mathbf{V}}(s)}{ds}. \tag{34}$$

3. We can calculate twist numbers for two neighboring magnetic field lines. Single out one field line and call it the 'axis'; neighboring field lines are almost parallel to this axis line but may spiral about it. Let $\mathbf{J} = \wedge\mathbf{B}/\mu_0$ be the electric current associated with the magnetic field $\mathbf{B}$, and $J_\| = \mathbf{J}\cdot\mathbf{B}/|\mathbf{B}|$. Then by Stokes' theorem applied to a small disc of radius r placed perpendicular to $\mathbf{B}$ on the axis,

$$\mu_0 \pi r^2 J_\| = 2\pi r B_\phi, \tag{35}$$

where B_ϕ is the azimuthal field component generated by the parallel current. The parallel magnetic field is B_s, which equals $|\mathbf{B}|$ on the axis. In a local cylindrical coordinate system (r, ϕ, s) surrounding the axis field line, we can ask how far a neighboring field line travels in both the ϕ and s directions. The field line equations relate this travel to the components of the field vector:

$$\frac{B_\phi}{B_s} = \frac{r\delta\phi}{\delta s}. \tag{36}$$

Putting these equations together gives

$$\frac{\mathrm{d}\mathcal{T}_w}{\mathrm{d}s} = \frac{1}{2\pi}\frac{\mathrm{d}\phi}{\mathrm{d}s} = \frac{\mu_0 J_\parallel}{4\pi|\mathbf{B}|}. \tag{37}$$

Similarly, if we measure how much two neighboring flow lines in a fluid twist about each other, then

$$\frac{\mathrm{d}\mathcal{T}_w}{\mathrm{d}s} = \frac{\omega_\parallel}{4\pi|\mathbf{V}|}, \tag{38}$$

where $\mathbf{V}$ is the fluid velocity and ω is the vorticity.

4. $\mathcal{T}_w$ is independent of the direction of the axis curve. This is because the neighboring curves must be almost aligned with the axis, so flipping the direction of the axis also flips the neighbors.
5. $\mathcal{T}_w$ equals half the average number of crossings between the two neighboring curves seen in projections of the curve onto a plane. The average is over all possible projection angles, and takes into account the sign of the crossing. Only local crossings are included in this average (if the axis curve crosses over some distant part of the secondary, this contributes to the writhe rather than the twist) [DH05].

Writhe also has some special properties:

1. Writhe depends on the axis curve only – the secondary curve is not needed for its definition.
2. Writhe equals the average number of crossings seen in projections of the curve. The average is over all possible projection angles [O94, DH05].
3. Writhe is independent of the direction of the axis curve.

4.2 Twisted Tubes

In many applications the picture of two curves twisting about each other is best replaced by thin tubes filled with curves. For example, vortex tubes and magnetic flux tubes play important roles in fluid mechanics and magneto-hydrodynamics. In engineering and material science, *isotropic rods* [VT00] are twisted tubes with some elastic energy function that is independent of positions in cross-sections of the tube. To obtain a twisted tube from an axis curve $\mathbf{x}(t)$, we first draw a circle around the axis at each position t. These circles should be of constant radius ϵ, and perpendicular to the axis (i.e. perpendicular to $\widehat{\mathbf{T}}(t)$). Joining up all the circles creates a tube enclosing the axis curve. We choose ϵ small enough so that the tube does not intersect itself.

Next we fill the tube with curves. These curves will be almost parallel to the axis, but may twist about the axis. We would like this twist to be uniform across a cross-section of the tube (but not necessarily uniform along the tube). Choose one of the curves and treat it like the secondary curve $\mathbf{y}$.

The tube surface can be given coordinates (t, ϕ) with $\phi = 0$ on $\mathbf{y}$, so $\mathbf{y}(t)$ has coordinates $(t, 0)$. Let $\widehat{\mathbf{W}} = \widehat{\mathbf{T}} \times \widehat{\mathbf{V}}$. Other curves (label them by the variable β) have different values of ϕ. Thus, on the surface (at radius ϵ)

$$\mathbf{y}(t, \beta) = \mathbf{x}(t) + \epsilon \left(\cos \beta \widehat{\mathbf{V}}(t) + \sin \beta \widehat{\mathbf{W}}(t) \right). \tag{39}$$

We can take the same tube with the same axis and fill it with many different sets of twisted curves. Each set, together with a particular choice of reference curve $\mathbf{y}$, defines a coordinate system in the tube. This coordinate system is called a *framing*. Some quantities, like the twist number between a secondary curve on the tube and the axis, explicitly depend on the choice of framing. Other quantities, like the writhe, depend only on the axis curve so are independent of framing.

We will sometimes need to take averages over all secondary curves. Given a number $f(\beta)$ describing the geometry of one of the secondary curves, let

$$\langle f \rangle \equiv \frac{1}{2\pi} \int_0^{2\pi} f(\beta) \, \mathrm{d}\beta. \tag{40}$$

5 Writhe from Winding Numbers

We can use the winding number techniques from sections 2 and 3 to simplify the calculation of writhe. These techniques will also help in extending the definition of writhe to open curves; in particular curves with endpoints on boundary planes (or spheres). Applications include isotropic rods fixed between two planes [S05, RM03] and magnetic arches in the solar atmosphere [TK05]. For full details of the calculations below, see [BP06].

Define the linking number $\widetilde{\mathcal{L}}(z_0)$ to be the net winding number between two curves between $z = -\infty$ and $z = z_0$. (One can replace $z = -\infty$ by $z_{\min}$, where $z_{\min}$ is the minimum height at which the two curves are both present.) Then the total linking number will be

$$\mathcal{L}_k = \widetilde{\mathcal{L}}(+\infty) = \widetilde{\mathcal{L}}(z_{\max}). \tag{41}$$

Also define $\widetilde{T}(z_0)$ and $\widetilde{\mathcal{W}}(z_0)$, the twist and writhe below z_0, in the same manner. We can then take derivatives in the z direction and ask how these quantities grow with height. We will see below that all secondary curves on a tube have the same growth in twist relative to the axis, i.e. $\mathrm{d}\widetilde{T}/\mathrm{d}z$ is independent of β. For linking number we need to be more careful; we will need to average over β.

Suppose we reorder the Călugăreanu formula as $\mathcal{W}_r = \mathcal{L}_k - \mathcal{T}_w$. We can then take this formula as a *definition* of writhe, rather deducing it from the standard definition equation (30). But first we must make sure that this formula makes sense, i.e. that it always gives the same answer for the axis curve, no matter which framing is used. For the differential writhe this vital property will be achieved below by averaging over β; thus we define

$$\frac{d\widetilde{\mathcal{W}}}{dz} \equiv \left\langle \frac{d\widetilde{\mathcal{L}}}{dz} - \frac{d\widetilde{\mathcal{T}}}{dz} \right\rangle. \tag{42}$$

To simplify the equations, we denote derivatives in z by a prime. Suppose the axis curve $\mathbf{x}$ has m intersections with the plane $z = $ constant; in other words m segments of the axis curve exist at height z. Each of these segments $\mathbf{x}_i$ has its corresponding secondary curve $\mathbf{y}_i$.

The linking number $\widetilde{\mathcal{L}}$ sums winding numbers over all pairs of segments $\mathbf{x}_i$ and $\mathbf{y}_j$. The axis curve segment $\mathbf{x}_i$ has secondary $\mathbf{y}_i$; if these wind about each other, they will contribute a *local* term

$$\widetilde{\mathcal{L}}_i = w_{ii} = w(\mathbf{x}_i, \mathbf{y}_i) \tag{43}$$

to the total linking number. In addition, axis segment i winds about far away secondary segments $\mathbf{y}_j$, $j \neq i$. These windings contribute *nonlocal* terms

$$\widetilde{\mathcal{L}}_{ij} \equiv \widetilde{\mathcal{L}}(\mathbf{x}_i, \mathbf{y}_j) \equiv w(\mathbf{x}_i, \mathbf{y}_j). \tag{44}$$

In sum, the z derivative decomposes into local (diagonal) terms, and nonlocal (off-diagonal) terms:

$$\widetilde{\mathcal{L}}'(z) = \sum_{i=1}^{n} \widetilde{\mathcal{L}}_i'(z) + \sum_{i=1}^{n} \sum_{\substack{j=1 \\ i \neq j}}^{n} \widetilde{\mathcal{L}}_{ij}'. \tag{45}$$

The following quantities will be useful:

$$\lambda = \frac{dz}{ds} = T_z; \tag{46}$$

$$\mu = |\hat{\mathbf{z}} \times \widehat{\mathbf{T}}| = \sqrt{1 - \lambda^2}. \tag{47}$$

5.1 The Twist as a Function of Height

For the moment, consider a segment (labelled i) of the axis curve which travels upwards, so that $\lambda > 0$. From equation (34) (modified considering equation (39)), the z derivative of the twist is

$$\widetilde{\mathcal{T}}_i'(z) = \frac{d\widetilde{\mathcal{T}}_i(z)}{dz} = \frac{d\mathcal{T}_{wi}}{ds} \frac{ds}{dz} \tag{48}$$

$$= \frac{1}{2\pi} \widehat{\mathbf{T}} \cdot \left(\cos \beta \widehat{\mathbf{V}} + \sin \beta \widehat{\mathbf{W}} \right) \times \left(\cos \beta \widehat{\mathbf{V}}' + \sin \beta \widehat{\mathbf{W}}' \right). \tag{49}$$

Because $\widehat{\mathbf{V}}$ and $\widehat{\mathbf{W}}$ are unit vectors $\widehat{\mathbf{V}} \cdot \widehat{\mathbf{V}}' = \widehat{\mathbf{W}} \cdot \widehat{\mathbf{W}}' = 0$. Also, $\widehat{\mathbf{V}} \cdot \widehat{\mathbf{W}} = 0$ so

$$\widehat{\mathbf{W}} \cdot \widehat{\mathbf{V}}' = -\widehat{\mathbf{V}} \cdot \widehat{\mathbf{W}}' \equiv \omega. \tag{50}$$

Thus

$$2\pi \widetilde{\mathcal{T}}_i' = (\cos \beta \, \widehat{\mathbf{W}} - \sin \beta \, \widehat{\mathbf{V}}) \cdot (\cos \beta \, \widehat{\mathbf{V}}' + \sin \beta \, \widehat{\mathbf{W}}') \tag{51}$$

$$= \omega. \tag{52}$$

The twist $\widetilde{\mathcal{T}}_i'$ is the same for all of the twisted curves on the surface of the tube, i.e. independent of β. However, ω does depend on the vectors $\widehat{\mathbf{V}}(z)$ and $\widehat{\mathbf{W}}(z)$. As we will see, this dependence on the framing will be cancelled when $\widetilde{\mathcal{T}}'$ is subtracted from $\widetilde{\mathcal{L}}'$ (averaged over β).

It will be convenient to define a new orthonormal frame:

$$\cdot \{\widehat{\mathbf{T}}, \widehat{\mathbf{f}}, \widehat{\mathbf{g}}\} = \{\widehat{\mathbf{T}}, \frac{\widehat{\mathbf{z}} \times \widehat{\mathbf{T}}}{\mu}, \widehat{\mathbf{T}} \times \frac{\widehat{\mathbf{z}} \times \widehat{\mathbf{T}}}{\mu}\}. \tag{53}$$

We can write $\widehat{\mathbf{V}}$ and $\widehat{\mathbf{W}}$ as combinations of $\widehat{\mathbf{f}}$ and $\widehat{\mathbf{g}}$ in terms of some angle $\psi(z)$:

$$\begin{pmatrix} \widehat{\mathbf{V}} \\ \widehat{\mathbf{W}} \end{pmatrix} = \begin{pmatrix} \cos \psi(z) & \sin \psi(z) \\ -\sin \psi(z) & \cos \psi(z) \end{pmatrix} \begin{pmatrix} \widehat{\mathbf{f}} \\ \widehat{\mathbf{g}} \end{pmatrix}. \tag{54}$$

The curve at position β is offset from the axis curve by

$$\widehat{\mathbf{U}} = \left(\cos \beta \widehat{\mathbf{V}} + \sin \beta \widehat{\mathbf{W}} \right) = \cos(\beta + \psi) \, \widehat{\mathbf{f}} + \sin(\beta + \psi) \, \widehat{\mathbf{g}}. \tag{55}$$

In terms of ψ one finds

$$2\pi \widetilde{\mathcal{T}}' = \omega = (\psi' + \widehat{\mathbf{f}}' \cdot \widehat{\mathbf{g}}) \tag{56}$$

$$= \psi' + \frac{\lambda}{\mu^2} \widehat{\mathbf{z}} \cdot \widehat{\mathbf{T}} \times \widehat{\mathbf{T}}'. \tag{57}$$

5.2 The Local Winding Number as a Function of Height

We now calculate the local winding between the axis and the secondary, as in equation (43):

$$\widetilde{\mathcal{L}}_i'(z) = w'(\mathbf{x}_i(z), \mathbf{y}_i(z)) = \frac{1}{2\pi}\Theta'(\mathbf{x}_i(z), \mathbf{y}_i(z)). \tag{58}$$

Here we run into a difficulty: the point on the secondary curve $\mathbf{y}_i(z)$ corresponds to a different arclength parameter s than the axis point $\mathbf{x}_i(z)$. In fact, to first order in ϵ (Berger & Prior 2006)

$$\mathbf{y}_i(z) - \mathbf{x}_i(z) = \epsilon\left(\widehat{\mathbf{U}} - \lambda^{-1}U_z\widehat{\mathbf{T}}\right). \tag{59}$$

Let $\mathbf{R}_i = (\mathbf{y}_i - \mathbf{x}_i)/\epsilon$. From equation (2),

$$2\pi\widetilde{\mathcal{L}}_i'(z) = \frac{\hat{\mathbf{z}}\cdot\mathbf{R}_i(z)\times\mathbf{R}_i'(z)}{|\mathbf{R}_i(z)|^2}. \tag{60}$$

After some algebra,

$$\frac{\hat{\mathbf{z}}\cdot\mathbf{R}_i\times\mathbf{R}_i'}{\mathbf{R}_i^2} = \frac{\lambda\psi' - \lambda'\cos(\beta+\psi)\sin(\beta+\psi)}{(\lambda^2\cos^2(\beta+\psi) + \sin^2(\beta+\psi))} + \frac{1}{\mu^2}\hat{\mathbf{z}}\cdot\widehat{\mathbf{T}}\times\widehat{\mathbf{T}}'. \tag{61}$$

This messy expression simplifies considerably if we average over all secondary curves in the tube:

$$2\pi\langle\widetilde{\mathcal{L}}_i'\rangle = \frac{1}{2\pi}\int_0^{2\pi}\frac{\hat{\mathbf{z}}\cdot\mathbf{R}_i\times\mathbf{R}_i'}{\mathbf{R}_i^2}\,\mathrm{d}\beta \tag{62}$$

$$= \psi' + \frac{1}{\mu^2}\hat{\mathbf{z}}\cdot\widehat{\mathbf{T}}\times\widehat{\mathbf{T}}'. \tag{63}$$

5.3 The Local Writhe as a Function of Height

We now, in effect, subtract section 5.1 from section 5.2. This will give us the local part of the writhe of segment i. From equation (57) and equation (63),

$$\widetilde{\mathcal{W}}_i' \equiv \langle\widetilde{\mathcal{L}}_i' - \widetilde{\mathcal{T}}_i'\rangle \tag{64}$$

$$= \frac{1-\lambda}{\mu^2}\hat{\mathbf{z}}\cdot\widehat{\mathbf{T}}\times\widehat{\mathbf{T}}'. \qquad (\lambda > 0). \tag{65}$$

So far we have assumed $\lambda = \mathrm{d}z/\mathrm{d}s > 0$ on segment i. We can obtain the result for negative λ most simply by noting that if we reverse arclength, i.e. let $s \to -s$, then linking, twist, and writhe do not change. Meanwhile, $\widehat{\mathbf{T}} \to -\widehat{\mathbf{T}}$ and $\widehat{\mathbf{T}}' \to -\widehat{\mathbf{T}}'$. Thus to preserve the value of $\widetilde{\mathcal{W}}_i'$, λ must be replaced by $|\lambda|$:

$$\widetilde{\mathcal{W}}_i' = \frac{1-|\lambda|}{\mu^2}\hat{\mathbf{z}}\cdot\widehat{\mathbf{T}}\times\widehat{\mathbf{T}}'. \tag{66}$$

(As an exercise, one may compute $\widetilde{W}_i'$ directly for negative λ. Some of the relations in the calculation above need amendment; for example if $s_{\max}$ is the maximum s value on the segment, then write $\widetilde{T}(z) = \mathcal{T}_w(s_{\max}) - \mathcal{T}_w(s(z))$. Also one must replace ψ by $\pi - \psi$.)

5.4 The Nonlocal Winding Number as a Function of Height

From equation (44), for $i \neq j$

$$\widetilde{\mathcal{L}}_{ij}'(z) = w'(\mathbf{x}_i(z), \mathbf{y}_j(z)). \tag{67}$$

But the distance between $\mathbf{x}_i(z)$ and $\mathbf{y}_j(z)$ is much greater than that between $\mathbf{x}_j(z)$ and $\mathbf{y}_j(z)$, so (strictly speaking, in the limit $\epsilon \to 0$)

$$\widetilde{\mathcal{L}}_{ij}'(z) = w'(\mathbf{x}_i(z), \mathbf{x}_j(z)) = \frac{1}{2\pi}\sigma_i\sigma_j\Theta'(\mathbf{x}_i(z), \mathbf{x}_j(z)) \qquad (i \neq j). \tag{68}$$

Thus the $m(m-1)$ off-diagonal terms in equation (45) depend only on the axis curve. These represent non-local windings of the different segments of the axis curve about each other.

We can now write down the entire expression for the writhe [BP06]:

Theorem 5.1.

$$\mathcal{W}_r = \widetilde{W}_{local} + \widetilde{W}_{nonlocal}; \tag{69}$$

$$\widetilde{W}_{local} = \frac{1}{2\pi}\sum_{i=1}^{n}\int_{z_i^{\min}}^{z_i^{\max}} \frac{1}{(1+|\lambda_i|)}(\widehat{\mathbf{T}}_i \times \widehat{\mathbf{T}}_i')_z \, dz; \tag{70}$$

$$\widetilde{W}_{nonlocal} = \sum_{i=1}^{n}\sum_{\substack{j=1\\i\neq j}}^{n}\frac{\sigma_i\sigma_j}{2\pi}\int_{z_{ij}^{\min}}^{z_{ij}^{\max}} \Theta_{ij}'(z) \, dz. \tag{71}$$

In terms of the intrinsic quantities *curvature* κ and *binormal* $\widehat{\mathbf{B}}$,

$$(\widehat{\mathbf{T}}_i \times \widehat{\mathbf{T}}_i')_z = \frac{\kappa_i B_{zi}}{\lambda_i}. \tag{72}$$

Note that the nonlocal writhe involves integrals of θ_{ij}'. We can treat these the same way as winding numbers, equation (17). In contrast to the linking number calculation, however, not all of the angles cancel. One finds that the orientations of the tangent vectors at the maxima and minima contribute to the sums. In particular, if $\phi(z_i^{\min})$ and $\phi(z_i^{\max})$ are the angles with respect to the x axis of the tangent vectors $\widehat{\mathbf{T}}(z_i^{\min})$ and $\widehat{\mathbf{T}}(z_i^{\max})$, then

$$\widetilde{\mathcal{W}}_{nonlocal} = \frac{1}{2\pi}\sum_{i=1}^{n}(\phi(z_i^{\min}) - \phi(z_i^{\max})) \quad -1 + \sum_{i=1}^{n}\sum_{\substack{j=1 \\ i\neq j}}^{n}\sigma_i\sigma_j n_{ij}. \tag{73}$$

Thus we do not need to perform the integrals completely – although we still need to keep track of the branch crossings to calculate the n_{ij} terms.

5.5 Example: A Trefoil Torus Knot

Figure 6 shows a trefoil knot. It is constructed by following a curve imbedded in a torus, which winds three times the short way around and two times the long way around. The 3 maxima and 3 minima in z divide the curve into 6 segments. For this curve the net local writhe of the 6 segments is -0.71, the nonlocal writhe between all pairs of segments sums to 4.23, giving a total writhe of $\mathcal{W}_r = 3.52$. Figure 7 displays a different version of the trefoil, a (3,2) torus knot; this curve has a higher writhe [MR92].

The trefoil figures also show the associated *tantrix curves*. As a tangent vector is defined to have unit norm, its tip lives on a unit sphere. As we go

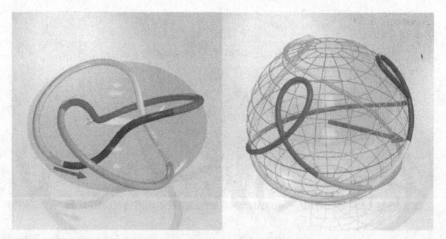

Fig. 6 The left figure displays a torus, with a ratio of major axis to minor axis equal to 2. A curve winds about this torus, forming a (2,3) torus knot, which is classified as a trefoil knot. The arrow shows the tangent vector at one point on the knot. Tangent vectors have unit norm, so if we place them inside a unit sphere, their tips just touch the surface. The figure on the right shows a unit sphere, called the *tantrix sphere*. The arrow touching this sphere represents the same tangent vector illustrated on the knot. The curve shown on the sphere gives the direction of the tangent vector at every point on the knot.

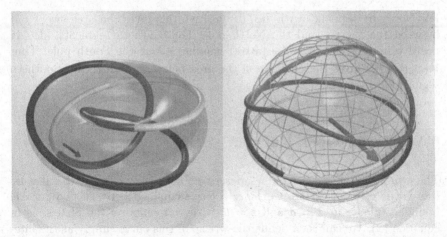

Fig. 7 As in figure 6, but with a (3,2) torus knot. For this curve, the local writhe is -0.51, the nonlocal writhe is 4.41, with a total writhe $\mathcal{W}_r = 3.90$.

around the knot, the tip of the tangent vector traces this curve on the *tantrix sphere*.

Recall that the knot is divided at extrema, where $T_z = \mathrm{d}z/\mathrm{d}s = 0$; this implies that the tantrix curve crosses the equator of the tantrix sphere at these points. The local writhe for each segment equals the area between the corresponding tantrix segment and the North pole (if the segment is in the Northern hemisphere) or the South pole (if in the Southern hemisphere). (The area enclosed by a curve drawn on a sphere is considered positive if the curve rotates anti-clockwise about the region and negative if clockwise.)

To see this, note that if θ gives latitude on the tantrix sphere, then $\lambda = \cos\theta$ and $\mu = \sin\theta$. Also,

$$\widehat{\mathbf{T}}'\,\mathrm{d}z = \mathrm{d}\widehat{\mathbf{T}} = \mathrm{d}\theta\,\hat{\theta} + \sin\theta\,\mathrm{d}\phi\,\hat{\phi}. \tag{74}$$

With $\hat{z} \times \widehat{\mathbf{T}} = \sin\theta\,\hat{\phi}$ we can write

$$(\widehat{\mathbf{T}} \times \widehat{\mathbf{T}}') \cdot \hat{z}\,\mathrm{d}z = \sin^2\theta\,\mathrm{d}\phi, \tag{75}$$

and so

$$2\pi\widetilde{\mathcal{W}}'_{local}(z) = \frac{1}{(1 + |\cos\theta|)}\sin^2\theta\frac{\mathrm{d}\phi}{\mathrm{d}z} \tag{76}$$

$$= (1 - |\cos\theta|)\frac{\mathrm{d}\phi}{\mathrm{d}z}. \tag{77}$$

Integrating between z_0 and z_1 gives

$$\widetilde{\mathcal{W}}_{local}(z_0, z_1) = \frac{1}{2\pi}\int_{z_0}^{z_1}(1 - |\cos\theta|)\frac{\mathrm{d}\phi}{\mathrm{d}z}\,\mathrm{d}z. \tag{78}$$

For Northern segments with $d\phi/dz$ positive this gives the area swept out between the tantrix and the North pole. For Southern segments, $d\phi/dz$ positive implies clockwise (negative) winding about the South pole. Thus $\widetilde{\mathcal{W}}_{local}(z_0, z_1)$ gives the negative of the area below the tantrix for Southern segments.

6 Writhe for Open Curves

The winding number techniques described here can be used to measure the writhe of open curves. Consider a curve confined between two parallel boundary planes at $z = a$ and $z = b$, with endpoints on the planes, as in equation (2). We can then define the writhe of this curve simply as the integral from a to b of $\widetilde{\mathcal{W}}'(z)$. This quantity is called the *polar writhe* in [BP06], as it is related to the area between the tantrix curve and the pole, as described in the previous section. Several other definitions appropriate for open writhe have appeared in the literature; these have usually involved considering perturbations of a reference curve [F78, AKT95, RM03] or closing the curve in a suitable way [S05].

The polar writhe is also relevant for *loops*, here defined as a curve confined to one half-space, with endpoints on the boundary plane of the half-space. Instead of a half-space and boundary plane, we can also consider loops extending from a boundary sphere (this requires modifying planar angles into spherical angles in the winding number calculations). Loops are of considerable importance in astrophysics, as magnetic field lines often form loops in the atmospheres of stars and accretion disks. Figure 8 shows one example from solar observations. Loops of magnetic flux often acquire a significant

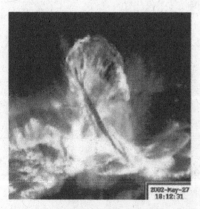 

Fig. 8 A filament of hot plasma in the solar corona observed by the TRACE detector on 27 May 2002. The plasma is thought to follow the path of magnetic field lines. The right picture displays a numerical simulation of the associated magnetic field [TK05].

Fig. 9 Two loops. For the short loop on the left, $\mathcal{W}_{r\,local} = -1.2$, $\mathcal{W}_{r\,nonlocal} = 1$. For the tall loop on the right, $\mathcal{W}_{r\,local} = 0.466$, $\mathcal{W}_{r\,nonlocal} = -2/3$. Thus the signs of both local and nonlocal writhe are opposite for the two loops, yet they have the same total polar writhe $\mathcal{W}_r = -0.2$. This demonstrates how the writhe can depend on height as well as which direction the loop turns. Note that in projection the tall loop has an S shape, while the short loop has an inverse S shape.

amount of twist; kink instabilities can convert some of this twist to writhe [LK97, Ba00, R05]).

The polar writhe is consistent with magnetic helicity integrals over volumes bounded by planes or spheres [BP06]). In particular, we can decompose the helicity of a magnetic flux loop of flux Φ as

$$H = (\mathcal{T}_w + \widetilde{\mathcal{W}})\Phi^2. \tag{79}$$

Figure 9 shows two loops; both were generated by taking an inverted parabola and twisting the top. The loops look quite different and yet have the same writhe. We can interpret this result in terms of the balance between local and nonlocal writhe. The tall loop has positive $\widetilde{\mathcal{W}}_{local}$ and negative $\widetilde{\mathcal{W}}_{nonlocal}$. The same total can be reached with negative $\mathcal{W}_{local}$ and positive $\mathcal{W}_{nonlocal}$, but with a shorter height. This may have implications for the interpretation of observations of structures in the solar atmosphere. Soft x-ray emissions in the shape of an S shaped curve, called *sigmoids*, are often seen in solar observations [RK96]. S shapes appear preferentially in the Southern hemisphere, whereas inverse S (or Z) shapes appear preferentially in the North. The shape may depend on the sign of writhe for the structure. However, assigning the sign of $\widetilde{\mathcal{W}}$ to an S or Z shape may depend on the height of the structure.

7 Higher Order Winding

Winding and Linking numbers do not capture all of the topological properties of curves. Figure 10 shows two braids. In the first braid, the curves wind about each other, but in a simple way. In the second braid, no two curves have net winding number, and yet the curves intertwine in a more complex way. This intertwining can be captured by computing a *second order winding number* [B91].

Suppose we replace Cartesian coordinates (x, y, z) by a pair (c, z) where c is a complex number $c = x + iy$. Let $(c_1(z), z)$, $(c_2(z), z)$ be two curves. Let

$$\lambda_{12}(z) = \frac{1}{2\pi i} \int_{-\infty}^{z} \sigma_1 \sigma_2 \frac{d \ln (c_2(z') - c_1(z'))}{dz'} \, dz'. \qquad (80)$$

Then the winding number is related to this complex logarithm, $w_{12}(z) = Re\, \lambda_{12}(z)$. Note that by defining λ_{12} as an integral of $d\ln(c_2 - c_1)/z$ rather than simply $\ln(c_2 - c_1)$, we allow winding numbers outside of the interval $(-1/2, 1/2)$.

Next, we can obtain second order winding numbers for three braided curves c_1, c_2, c_3 by integrating suitable combinations of the λ functions. Suppose the curves travel from $z = 0$ to $z = 1$. Then the second order winding number is

$$\Psi = \tfrac{1}{2} \int_0^1 \left((\lambda_{12} - \lambda_{23}) \, d\lambda_{31} + (\lambda_{23} - \lambda_{31}) \, d\lambda_{12} + (\lambda_{31} - \lambda_{12}) \, d\lambda_{23} \right). \qquad (81)$$

This quantity is invariant to deformations of the curves, leaving their endpoints fixed. Third and higher order invariants for braids [B01] or knots and links [CD00] require the machinery of Kontsevich integrals [K93] to identify the appropriate combinations of λ functions.

Fig. 10 Two braids. The left braid only exhibits ordinary (first order) winding. The right figure shows a pigtail braid with second order winding.

References

[AKT95]   Aldinger J, Klapper I, & Tabor M: Formulae for the calculation and estimation of writhe. J. Knot Theory Ram., **4**, 343–372 (1995)

[B91]   Berger M A: Third order braid invariants. J. Physics A: Mathematical and General, **24**, 4027–4036 (1991)

[B01]   Berger M A: Topological invariants in braid theory. Letters in Math. Physics, **55**, 181–192 (2001)

[BP06]   Berger M A & Prior P: The writhe of open and closed curves. J. Physics A: Mathematical and General, **39**, 8321–8348 (2006)

[Ba00]   Baty H: Magnetic topology during the reconnection process in a kinked coronal loop. Astronomy and Astrophysics, **360**, 345–350 (2000)

[C59]   Călugăreanu G: Sur les classes d'isotopie des noeuds tridimensionnels et leurs invariants. Czechoslovak Math J, **11**, 588–625 (1959)

[C05]   Cantarella J: On comparing the writhe of a smooth curve to the writhe of an inscribed polygon. SIAM J. of Numerical Analysis, **42**, 1846–1861 (2005)

[CD00]   Chmutov S V& Duzhin S V: The Kontsevich Integral. Acta Appl. Math., **66**, 155–190 (2000)

[DH05]   Dennis M R & Hannay J H: Geometry of Călugăreanu 's theorem. Proc. Roy. Soc. A, **461**, 3245-3254 (2005)

[F78]   Fuller F B: Decomposition of the linking of a ribbon: a problem from molecular biology. Proc. Natl. Acad. Sci. USA, **75**, 3557–3561 (1978)

[GVV03]   Ghrist R W, Van den Berg J B, & Vandervorst R C: Morse theory on spaces of braids and Lagrangian dynamics. Inventiones Mathematicae, **152**, 369–432 (2003)

[K93]   Kontsevich M: Vassiliev's knot invariants. Adv. Soviet Math., **16**, 137–150 (1993)

[KHS06]   Kristiansen K D, Helgesen G, Skjeltorp A T: Braid theory and Zipf-Mandelbrot relation used in microparticle dynamics. European Physical J., **B 51**, 363–371 (2006)

[LK97]   Longcope D W & Klapper I: Dynamics of a thin twisted flux tube. Astrophysical J., **488**, 443–453 (1997)

[MR92]   Moffatt H K & Ricca R L: Helicity and the Călugăreanu invariant. Proc. Roy. Soc. A, **439**, 411–429 (1992)

[O94]   Orlandini E, Test M C, Whittington S G, Sumners D W, & Janse van Rensburg E J: The writhe of a self-avoiding walk. J. Physics A: Mathematical and General, **27**, L333–L338 (1994)

[R05]   Ricca R L: Inflexional disequilibrium of magnetic flux-tubes. Fluid Dynamics Research, **36**, 319–332 (2005)

[RM03]   Rossetto V & Maggs A C: Writhing geometry of Open DNA. J. Chem. Phys., **118**, 9864–9874 (2003)

[RK96]   Rust D M & Kumar A: Evidence for helically kinked magnetic flux ropes in solar eruptions. Astrophys. J. Lett., **464**, L199-L202 (1996)

[S05]   Starostin E L: On the writhing number of a non-closed curve. In: Calvo J, Millett K, Rawdon E, & Stasiak A (eds) Physical and Numerical Models in Knot Theory Including Applications to the Life Sciences. Series on Knots and Everything, World Scientific Publishing, Singapore 525-545 (2005)

[TK05]   Török T & Kliem B: Confined and ejective eruptions of kink-unstable flux ropes. Astrophysical J., **630**, L97–L100 (2005)

[VT00]   van der Heijden G H M & Thompson J M T: Helical and localised buckling in twisted rods: A unified analysis of the symmetric case. Nonlinear Dynamics, **21**, 71–99 (2000)

[VM97]   Vologodskii A V & Marko J F: Extension of torsionally stressed DNA by external force. Biophys. J., **73**, 123–132 (1997)

Tangles, Rational Knots and DNA

Louis H. Kauffman (CIME Lecturer) and Sofia Lambropoulou

Abstract This paper draws a line from the basics of rational tangles to the tangle model of DNA recombination. We sketch the classification of rational tangles, unoriented and oriented rational knots and the application of these subjects to DNA recombination.

1 Introduction

Rational knots and links are a class of alternating links of one or two unknotted components, and they are the easiest knots to make (also for Nature!). The first twenty five knots, except for 8_5, are rational. Furthermore all knots and links up to ten crossings are either rational or are obtained from rational knots by insertion operations on certain simple graphs. Rational knots are also known in the literature as four-plats, Viergeflechte and 2-bridge knots. The lens spaces arise as 2-fold branched coverings along rational knots.

A rational tangle is the result of consecutive twists on neighbouring endpoints of two trivial arcs, see Definition 1. Rational knots are obtained by taking numerator closures of rational tangles (see Figure 19), which form a basis for their classification. Rational knots and rational tangles are of fundamental importance in the study of DNA recombination. Rational knots and

L.H. Kauffman
Department of Mathematics and Computer Science, M/C 249
University of Illinois at Chicago 851 S Morgan St. Chicago, IL 60607-7045, USA
e-mail: kauffman@math.uic.edu
http://www.math.uic.edu/~kauffman

S. Lambropoulou
Department of Mathematics National Technical University
Zografou Campus GR-15780 Athens, GREECE
e-mail: sofia@math.ntua.gr
http://users.ntua.gr/sofial

R.L. Ricca (ed.), *Lectures on Topological Fluid Mechanics,*
Lecture Notes in Mathematics 1973, DOI: 10.1007/978-3-642-00837-5,
© Springer-Verlag Berlin Heidelberg 2009

links were first considered in [41] and [2]. Treatments of various aspects of
rational knots and rational tangles can be found in [3], [7], [47], [6], [43], [17],
[28], [32], [35]. A rational tangle is associated in a canonical manner with
a unique, reduced rational number or ∞, called *the fraction* of the tangle.
Rational tangles are classified by their fractions by means of the following
theorem:

Theorem 1.1 (Conway, 1970). *Two rational tangles are isotopic if and
only if they have the same fraction.*

John H. Conway [7] introduced the notion of tangle and defined the fraction of a rational tangle using the continued fraction form of the tangle
and the Alexander polynomial of knots. Via the Alexander polynomial, the
fraction is defined for the larger class of all 2-tangles. In this paper we are
interested in different definitions of the fraction, and we give a self-contained
exposition of the construction of the invariant fraction for arbitrary 2-tangles
from the bracket polynomial [20]. The tangle fraction is a key ingredient in
both the classification of rational knots and in the applications of knot theory
to DNA. Proofs of Theorem 1 can be found in [34], [6] p.196, [17] and [26].

More than one rational tangle can yield the same or isotopic rational knots
and the equivalence relation between the rational tangles is reflected in an
arithmetic equivalence of their corresponding fractions. This is marked by a
theorem due originally to Schubert [46] and reformulated by Conway [7] in
terms of rational tangles.

Theorem 1.2 (Schubert, 1956). *Suppose that rational tangles with frac-
tions $\frac{p}{q}$ and $\frac{p'}{q'}$ are given (p and q are relatively prime. Similarly for p' and q'.)
If $K(\frac{p}{q})$ and $K(\frac{p'}{q'})$ denote the corresponding rational knots obtained by taking
numerator closures of these tangles, then $K(\frac{p}{q})$ and $K(\frac{p'}{q'})$ are topologically
equivalent if and only if*

1. $p = p'$ and
2. either $q \equiv q' (mod\, p)$ or $qq' \equiv 1 (mod\, p)$.

This classic theorem [46] was originally proved by using an observation
of Seifert that the 2-fold branched covering spaces of S^3 along $K(\frac{p}{q})$ and
$K(\frac{p'}{q'})$ are lens spaces, and invoking the results of Reidemeister [42] on the
classification of lens spaces. Another proof using covering spaces has been
given by Burde in [5]. Schubert also extended this theorem to the case of
oriented rational knots and links described as 2-bridge links:

Theorem 1.3 (Schubert, 1956). *Suppose that orientation-compatible
rational tangles with fractions $\frac{p}{q}$ and $\frac{p'}{q'}$ are given with q and q' odd. (p and q
are relatively prime. Similarly for p' and q'.) If $K(\frac{p}{q})$ and $K(\frac{p'}{q'})$ denote the
corresponding rational knots obtained by taking numerator closures of these
tangles, then $K(\frac{p}{q})$ and $K(\frac{p'}{q'})$ are topologically equivalent if and only if*

1. $p = p'$ and
2. either $q \equiv q'(mod\,2p)$ or $qq' \equiv 1(mod\,2p)$.

In [27] we give the first combinatorial proofs of Theorem 2 and Theorem 3. In this paper we sketch the proofs in [26] and [27] of the above three theorems and we give the key examples that are behind all of our proofs. We also give some applications of Theorems 2 and 3 using our methods.

The paper is organized as follows. In Section 2 we introduce 2-tangles and rational tangles, Reideimeister moves, isotopies and operations. We give the definition of flyping, and state the (now proved) Tait flyping conjecture. The Tait conjecture is used implicitly in our classification work. In Section 3 we introduce the continued fraction expression for rational tangles and its properties. We use the continued fraction expression for rational tangles to define their fractions. Then rational tangle diagrams are shown to be isotopic to alternating diagrams. The alternating form is used to obtain a canonical form for rational tangles, and we obtain a proof of Theorem 1.

Section 4 discusses alternate definitions of the tangle fraction. We begin with a self-contained exposition of the bracket polynomial for knots, links and tangles. Using the bracket polynomial we'define a fraction $F(T)$ for arbitrary 2-tangles and show that it has a list of properties that are sufficient to prove that for T rational, $F(T)$ is identical to the continued fraction value of T, as defined in Section 3. The next part of Section 4 gives a different definition of the fraction of a rational tangle, based on coloring the tangle arcs with integers. This definition is restricted to rational tangles and those tangles that are obtained from them by tangle-arithmetic operations, but it is truly elementary, depending just on a little algebra and the properties of the Reidemeister moves. Finally, we sketch yet another definition of the fraction for 2-tangles that shows it to be the value of the conductance of an electrical network associated with the tangle.

Section 5 contains a description of our approach to the proof of Theorem 2, the classification of unoriented rational knots and links. The key to this approach is enumerating the different rational tangles whose numerator closure is a given unoriented rational knot or link, and confirming that the corresponding fractions of these tangles satisfy the arithmetic relations of the Theorem. Section 6 sketches the classification of rational knots and links that are isotopic to their mirror images. Such links are all closures of palindromic continued fraction forms of even length. Section 7 describes our proof of Theorem 3, the classification of oriented rational knots. The statement of Theorem 3 differs from the statement of Theorem 2 in the use of integers modulo $2p$ rather than p. We see how this difference arises in relation to matching orientations on tangles. This section also includes an explanation of the fact that fractions with even numerators correspond to rational links of two components, while fractions with odd numerators correspond to single component rational knots (the denominators are odd in both cases). Section 8 discusses strongly invertible rational knots and links. These correspond to palindromic continued fractions of odd length.

Section 9 is an introduction to the tangle model for DNA recombination. The classification of the rational knots and links, and the use of the tangle fractions is the basic topology behind the tangle model for DNA recombination. We indicate how problems in this model are reduced to properties of rational knots, links and tangles, and we show how a finite number of observations of successive DNA recombination can pinpoint the recombination mechanism.

2 2-Tangles and Rational Tangles

Throughout this paper we will be working with 2-*tangles*. The theory of tangles was discovered by John Conway [7] in his work on enumerating and classifying knots. A 2-tangle is an embedding of two arcs (homeomorphic to the interval [0,1]) and circles into a three-dimensional ball B^3 standardly embedded in Euclidean three-space S^3, such that the endpoints of the arcs go to a specific set of four points on the surface of the ball, so that the circles and the interiors of the arcs are embedded in the interior of the ball. The left-hand side of Figure 1 illustrates a 2-tangle. Finally, a 2-tangle is *oriented* if we assign orientations to each arc and each circle. Without loss of generality, the four endpoints of a 2-tangle can be arranged on a great circle on the boundary of the ball. One can then define a *diagram* of a 2-tangle to be a regular projection of the tangle on the plane of this great circle. In illustrations we may replace this circle by a box.

The simplest possible 2-tangles comprise two unlinked arcs either horizontal or vertical. These are the *trivial tangles*, denoted [0] and [∞] tangles respectively, see Figure 2.

Definition 2.1. A 2-tangle is *rational* if it can be obtained by applying a finite number of consecutive twists of neighbouring endpoints to the elementary tangles [0] or [∞].

The simplest rational tangles are the [0], the [∞], the [+1] and the [−1] tangles, as illustrated in Figure 3, while the next simplest ones are:

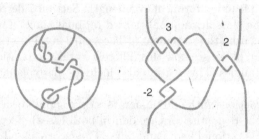

Fig. 1 A 2-tangle and a rational tangle.

Fig. 2 The trivial tangles
[0] and [∞].

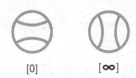

[0]   [∞]

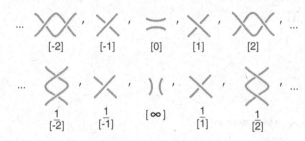

[-2]   [-1]   [0]   [1]   [2]

$\frac{1}{[-2]}$ $\frac{1}{[-1]}$ [∞] $\frac{1}{[1]}$ $\frac{1}{[2]}$

Fig. 3 The elementary rational tangles.

(i) The *integer tangles*, denoted by $[n]$, made of n horizontal twists, $n \in \mathbb{Z}$.
(ii) The *vertical tangles*, denoted by $\frac{1}{[n]}$, made of n vertical twists, $n \in \mathbb{Z}$.

These are the inverses of the integer tangles, see Figure 3. This terminology will be clear soon.

Examples of rational tangles are illustrated in the right-hand side of Figure 1 as well as in Figures 8 and 17 below.

We study tangles up to *isotopy*. Two 2-tangles, T, S, in B^3 are said to be *isotopic*, denoted by $T \sim S$, if they have identical configurations of their four endpoints in the boundary S^2 of the three-ball, and there is an ambient isotopy of (B^3, T) to (B^3, S) that is the identity on the boundary $(S^2, \partial T) = (S^2, \partial S)$. An ambient isotopy can be imagined as a continuous deformation of B^3 fixing the four endpoints on the boundary sphere, and bringing one tangle to the other without causing any self-intersections.

In terms of diagrams, Reidemeister [40] proved that the local moves on diagrams illustrated in Figure 4 capture combinatorially the notion of ambient isotopy of knots, links and tangles in three-dimensional space. That is, if two diagrams represent knots, links or tangles that are isotopic, then the one diagram can be obtained from the other by a sequence of *Reidemeister moves*. In the case of tangles *the endpoints of the tangle remain fixed* and all the moves occur inside the tangle box.

Two oriented 2-tangles are said to be *oriented isotopic* if there is an isotopy between them that preserves the orientations of the corresponding arcs and the corresponding circles. The diagrams of two oriented isotopic tangles differ by a sequence of *oriented Reidemeister moves*, i.e. Reidemeister moves with orientations on the little arcs that remain consistent during the moves.

From now on we will be thinking in terms of tangle diagrams. Also, we will be referring to both knots and links whenever we say 'knots'.

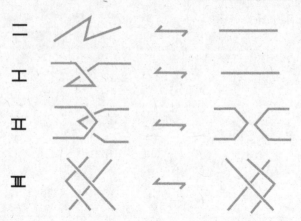

Fig. 4 The Reidemeister moves.

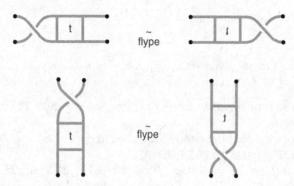

Fig. 5 The flype moves.

A *flype* is an isotopy move applied on a 2-subtangle of a larger tangle
or knot as shown in Figure 5. A flype preserves the alternating structure
of a diagram. Even more, flypes are the only isotopy moves needed in the
statement of the celebrated Tait Conjecture for alternating knots, stating
that *two alternating knots are isotopic if and only if any two corresponding
diagrams on S^2 are related by a finite sequence of flypes.* This was posed by
P.G. Tait, [50] in 1898 and was proved by W. Menasco and M. Thistlethwaite,
[33] in 1993.

The class of 2-tangles is closed under the operations of *addition* (+) and
multiplication (∗) as illustrated in Figure 6. Adddition is accomplished by
placing the tangles side-by-side and attaching the NE strand of the left tangle
to the NW strand of the right tangle, while attaching the SE strand of the
left tangle to the SW strand of the right tangle. The product is accomplished
by placing one tangle underneath the other and attaching the upper strands
of the lower tangle to the lower strands of the upper tangle.

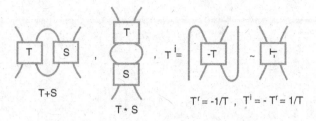

Fig. 6 Addition, product and inversion of 2-tangles.

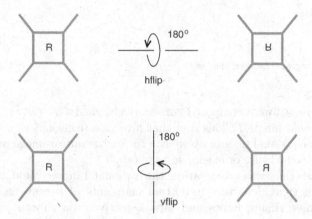

Fig. 7 The horizontal and the vertical flip.

The *mirror image* of a tangle T is denoted by $-T$ and it is obtained by switching all the crossings in T. Another operation is *rotation* accomplished by turning the tangle counter-clockwise by $90°$ in the plane. The rotation of T is denoted by T^r. The *inverse* of a tangle T, denoted by $1/T$, is defined to be $-T^r$. See Figure 6. In general, the inversion or rotation of a 2-tangle is an order 4 operation. Remarkably, for rational tangles the inversion (rotation) is an order 2 operation. It is for this reason that we denote the inverse of a 2-tangle T by $1/T$ or T^{-1}, and hence the rotate of the tangle T can be denoted by $-1/T = -T^{-1}$.

We describe now another operation applied on 2-tangles, which turns out to be an isotopy on rational tangles. We say that R^{hflip} is the *horizontal flip* of the tangle R if R^{hflip} is obtained from R by a $180°$ rotation around a horizontal axis on the plane of R. Moreover, R^{vflip} is the *vertical flip* of the 2-tangle R if R^{vflip} is obtained from R by a $180°$ rotation around a vertical axis on the plane of R. See Figure 7 for illustrations. Note that a flip switches the endpoints of the tangle and, in general, a flipped tangle is not isotopic to the original one. *It is a property of rational tangles that $T \sim T^{hflip}$ and $T \sim T^{vflip}$ for any rational tangle T.* This is obvious for the tangles $[n]$ and $\frac{1}{[n]}$. The general proof crucially uses flypes, see [26].

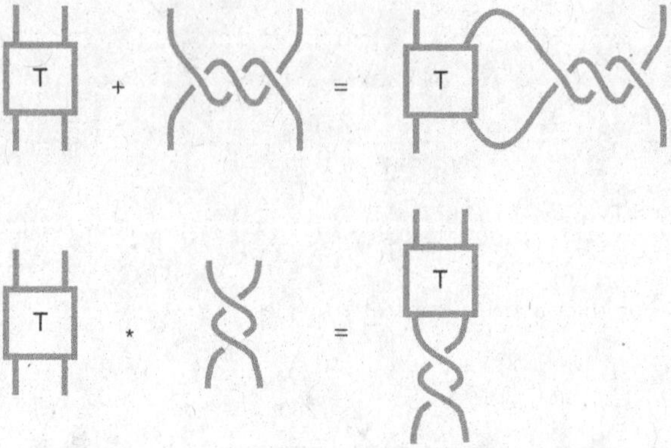

Fig. 8 Creating new rational tangles.

The above isotopies composed consecutively yield $T \sim (T^{-1})^{-1} = (T^r)^r$ for any rational tangle T. This says that inversion (rotation) is an operation of order 2 for rational tangles, so we can rotate the mirror image of T by 90° either counterclockwise or clockwise to obtain T^{-1}.

Note that the twists generating the rational tangles could take place between the right, left, top or bottom endpoints of a previously created rational tangle. Using flypes and flips inductively on subtangles one can always bring the twists to the right or bottom of the rational tangle. We shall then say that the rational tangle is in *standard form*. Thus a rational tangle in standard form is created by consecutive additions of the tangles [±1] *only on the right* and multiplications by the tangles [±1] *only at the bottom*, starting from the tangles [0] or [∞]. For example, Figure 1 illustrates the tangle $(([3] * \frac{1}{[-2]}) + [2])$, while Figure 17 illustrates the tangle $(([3] * \frac{1}{[2]}) + [2])$ in standard form. Figure 8 illustrates addition on the right and multiplication on the bottom by elementary tangles.

We also have the following *closing* operations, which yield two different knots: the *Numerator* of a 2-tangle T, denoted by $N(T)$, obtained by joining with simple arcs the two upper endpoints and the two lower endpoints of T, and the *Denominator* of a 2-tangle T, obtained by joining with simple arcs each pair of the corresponding top and bottom endpoints of T, denoted by $D(T)$. We have $N(T) = D(T^r)$ and $D(T) = N(T^r)$. We note that every knot or link can be regarded as the numerator closure of a 2-tangle.

We obtain $D(T)$ from $N(T)$ by a [0] − [∞] interchange, as shown in Figure 10. This 'transmutation' of the numerator to the denominator is a precursor to the tangle model of a recombination event in DNA, see Section 9. The [0] − [∞] interchange can be described algebraically by the equations:

$$N(T) = N(T + [0]) \longrightarrow N(T + [\infty]) = D(T).$$

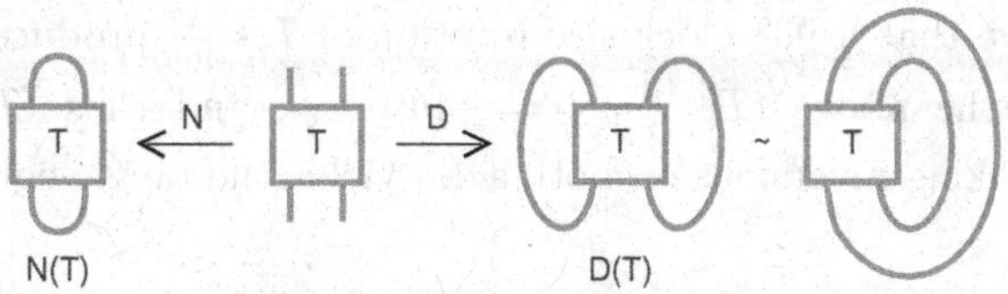

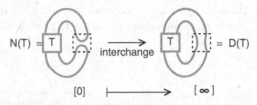

Fig. 9 The numerator and denominator of a 2-tangle.

$$N(T) = \boxed{T}\ \xrightarrow{\text{interchange}}\ \boxed{T} = D(T)$$

$$[0]\ \longmapsto\ [\infty]$$

Fig. 10 The $[0] - [\infty]$ interchange.

We will concentrate on the class of *rational knots and links* arising from closing the rational tangles. Even though the sum/product of rational tangles is in general not rational, the numerator (denominator) closure of the sum/product of two rational tangles is still a rational knot. It may happen that two rational tangles are not isotopic but have isotopic numerators. This is the basic idea behind the classification of rational knots, see Section 5.

3 Continued Fractions and the Classification of Rational Tangles

In this section we assign to a rational tangle a fraction, and we explore the analogy between rational tangles and continued fractions. This analogy culminates in a common canonical form, which is used to deduce the classification of rational tangles.

We first observe that multiplication of a rational tangle T by $\frac{1}{[n]}$ may be obtained as addition of $[n]$ to the inverse $\frac{1}{T}$ followed by inversion. Indeed, we have:

Lemma 3.1. *The following tangle equation holds for any rational tangle T.*

$$T * \frac{1}{[n]} = \frac{1}{[n] + \frac{1}{T}}.$$

Thus any rational tangle can be built by a series of the following operations: *Addition of* $[\pm 1]$ *and Inversion.*

Proof. Observe that a 90° clockwise rotation of $T * \frac{1}{[n]}$ produces $-[n] - \frac{1}{T}$. Hence, from the above $(T * \frac{1}{[n]})^r = -[n] - \frac{1}{T}$, and thus $(T * \frac{1}{[n]})^{-1} = [n] + \frac{1}{T}$. So, taking inversions on both sides yields the tangle equation of the statement. $\square$

Definition 3.2. A *continued fraction in integer tangles* is an algebraic description of a rational tangle via a continued fraction built from the tangles $[a_1], [a_2], \ldots, [a_n]$ with all numerators equal to 1, namely an expression of the type:

$$[[a_1], [a_2], \ldots, [a_n]] := [a_1] + \cfrac{1}{[a_2] + \cdots + \cfrac{1}{[a_{n-1}] + \frac{1}{[a_n]}}}$$

for $a_2, \ldots, a_n \in \mathbb{Z} - \{0\}$ and n even or odd. We allow that the term a_1 may be zero, and in this case the tangle $[0]$ may be omitted. A rational tangle described via a continued fraction in integer tangles is said to be in *continued fraction form*. The *length* of the continued fraction is arbitrary – in the previous formula illustrated with length n – whether the first summand is the tangle $[0]$ or not.

It follows from Lemma 3.2 that inductively *every rational tangle can be written in continued fraction form*. Lemma 3.2 makes it easy to write out the continued fraction form of a given rational tangle, since horizontal twists are integer additions, and multiplications by vertical twists are the reciprocals of integer additions. For example, Figure 1 illustrates the rational tangle $[2] + \frac{1}{[-2] + \frac{1}{[3]}}$, Figure 17 illustrates the rational tangle $[2] + \frac{1}{[2] + \frac{1}{[3]}}$. Note that $([c] * \frac{1}{[b]}) + [a]$ has the continued fraction form $[a] + \cfrac{1}{[b] + \frac{1}{[c]}} = [[a], [b], [c]]$. For $T = [[a_1], [a_2], \ldots, [a_n]]$ the following statements are now straightforward.

1. $T + [\pm 1] = [[a_1 \pm 1], [a_2], \ldots, [a_n]]$,

2. $\quad \frac{1}{T} = [[0], [a_1], [a_2], \ldots, [a_n]]$,

3. $\quad -T = [[-a_1], [-a_2], \ldots, [-a_n]]$.

We now recall some facts about continued fractions. See for example [29], [36], [30], [51]. In this paper we shall only consider continued fractions of the type

$$[a_1, a_2, \ldots, a_n] := a_1 + \cfrac{1}{a_2 + \cdots + \cfrac{1}{a_{n-1} + \frac{1}{a_n}}}$$

for $a_1 \in \mathbb{Z}$, $a_2, \ldots, a_n \in \mathbb{Z} - \{0\}$ and n even or odd. The *length* of the continued fraction is the number n whether a_1 is zero or not. Note that if for $i > 1$ all terms are positive or all terms are negative and $a_1 \neq 0$

($a_1 = 0$,) then the absolute value of the continued fraction is greater (smaller) than one. Clearly, the two simple algebraic operations *addition of $+1$ or -1* and *inversion* generate inductively the whole class of continued fractions starting from zero. For any rational number $\frac{p}{q}$ the following statements are straightforward.

1. there are $a_1 \in \mathbb{Z}$, $a_2, \ldots, a_n \in \mathbb{Z} - \{0\}$ such that $\frac{p}{q} = [a_1, a_2, \ldots, a_n]$,
2. $\frac{p}{q} \pm 1 = [a_1 \pm 1, a_2, \ldots, a_n]$,
3. $\qquad \frac{q}{p} = [0, a_1, a_2, \ldots, a_n]$,
4. $\qquad -\frac{p}{q} = [-a_1, -a_2, \ldots, -a_n]$.

We can now define the fraction of a rational tangle.

Definition 3.3. Let T be a rational tangle isotopic to the continued fraction form $[[a_1], [a_2], \ldots, [a_n]]$. We define *the fraction $F(T)$ of T* to be the numerical value of the continued fraction obtained by substituting integers for the integer tangles in the expression for T, i.e.

$$F(T) := a_1 + \cfrac{1}{a_2 + \cdots + \cfrac{1}{a_{n-1} + \frac{1}{a_n}}} = [a_1, a_2, \ldots, a_n],$$

if $T \neq [\infty]$, and $F([\infty]) := \infty = \frac{1}{0}$, as a formal expression.

Remark 3.4. This definition is good in the sense that one can show that isotopic rational tangles always differ by flypes, and that the fraction is unchanged by flypes [26].

Clearly the tangle fraction has the following properties.

1. $F(T + [\pm 1]) = F(T) \pm 1$,
2. $\qquad F(\frac{1}{T}) = \frac{1}{F(T)}$,
3. $\qquad F(-T) = -F(T)$.

The main result about rational tangles (Theorem 1) is that two rational tangles are isotopic if and only if they have the same fraction. We will show that every rational tangle is isotopic to a unique alternating continued fraction form, and that this alternating form can be deduced from the fraction of the tangle. The Theorem then follows from this observation.

Lemma 3.5. *Every rational tangle is isotopic to an alternating rational tangle.*

Proof. Indeed, if T has a non-alternating continued fraction form then the following configuration, shown in the left of Figure 11, must occur somewhere in T, corresponding to a change of sign from one term to an adjacent term

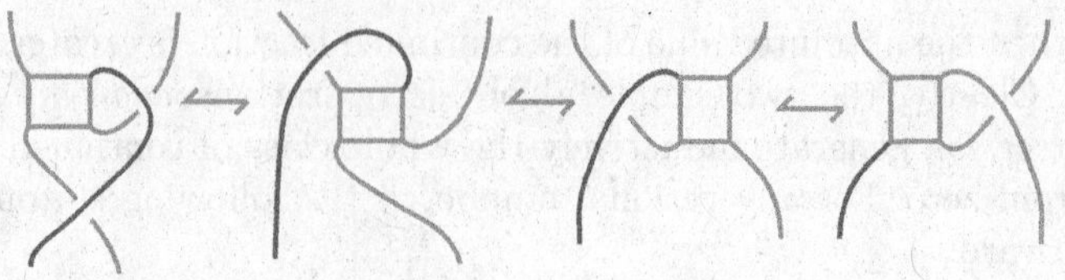

Fig. 11 Reducing to the alternating form.

in the tangle continued fraction. This configuration is isotopic to a simpler isotopic configuration as shown in that figure.

Therefore, it follows by induction on the number of crossings in the tangle that T is isotopic to an alternating rational tangle. $\qquad\square$

Recall that a tangle is alternating if and only if it has crossings all of the same type. Thus, *a rational tangle $T = [[a_1], [a_2], \ldots, [a_n]]$ is alternating if the a_i's are all positive or all negative.* For example, the tangle of Figure 17 is alternating.

A rational tangle $T = [[a_1], [a_2], \ldots, [a_n]]$ is said to be in *canonical form* if T is alternating and n is odd. The tangle of Figure 17 is in canonical form. We note that if T is alternating and n even, then we can bring T to canonical form by breaking a_n by a unit, e.g. $[[a_1], [a_2], \ldots, [a_n]] = [[a_1], [a_2], \ldots, [a_n - 1], [1]]$, if $a_n > 0$.

The last key observation is the following well-known fact about continued fractions.

Lemma 3.6. *Every continued fraction $[a_1, a_2, \ldots, a_n]$ can be transformed to a unique canonical form $[\beta_1, \beta_2, \ldots, \beta_m]$, where all β_i's are positive or all negative integers and m is odd.*

Proof. It follows immediately from Euclid's algorithm. We evaluate first $[a_1, a_2, \ldots, a_n] = \frac{p}{q}$, and using Euclid's algorithm we rewrite $\frac{p}{q}$ in the desired form. We illustrate the proof with an example. Suppose that $\frac{p}{q} = \frac{11}{7}$. Then

$$\frac{11}{7} = 1 + \frac{4}{7} = 1 + \frac{1}{\frac{7}{4}} = 1 + \frac{1}{1 + \frac{3}{4}} = 1 + \frac{1}{1 + \frac{1}{\frac{4}{3}}}$$
$$= 1 + \frac{1}{1 + \frac{1}{1 + \frac{1}{3}}} = [1, 1, 1, 3] = 1 + \frac{1}{1 + \frac{1}{1 + \frac{1}{2 + \frac{1}{1}}}} = [1, 1, 1, 2, 1].$$

This completes the proof. $\qquad\square$

Note that if $T = [[a_1], [a_2], \ldots, [a_n]]$ and $S = [[b_1], [b_2], \ldots, [b_m]]$ are rational tangles in canonical form with the same fraction, then it follows from this Lemma that $[a_1, a_2, \ldots, a_n]$ and $[b_1, b_2, \ldots, b_m]$ are canonical continued fraction forms for the same rational number, and hence are equal term-by-term. Thus the uniqueness of canonical forms for continued fractions implies the uniqueness of canonical forms for rational tangles. For example, let $T = [[2], [-3], [5]]$. Then $F(T) = [2, -3, 5] = \frac{23}{14}$. But $\frac{23}{14} = [1, 1, 1, 1, 1, 4]$, thus $T \sim [[1], [1], [1], [1], [4]]$, and this last tangle is the canonical form of T.

Proof of Theorem 1. We have now assembled all the ingredients for the proof of Theorem 1. In one direction, suppose that rational tangles T and S are isotopic. Then each is isotopic to its canonical form T' and S' by a sequence of flypes. Hence the alternating tangles T' and S' are isotopic to one another. By the Tait conjecture, there is a sequence of flypes from T' to S'. Hence there is a sequence of flypes from T to S. One verifies that the fraction as we defined it is invariant under flypes. Hence T and S have the same fraction. In the other direction, suppose that T and S have the same fraction. Then, by the remark above, they have identical canonical forms to which they are isotopic, and therefore they are isotopic to each other. This completes the proof of the Theorem. □

4 Alternate Definitions of the Tangle Fraction

In the last section and in [26] the fraction of a rational tangle is defined directly from its combinatorial structure, and we verify the topological invariance of the fraction using the Tait conjecture.

In [26] we give yet another definition of the fraction for rational tangles by using coloring of the tangle arcs. There are definitions that associate a fraction $F(T)$ (including $0/1$ and $1/0$) to any 2-tangle T whether or not it is rational. The first definition is due to John Conway in [7] using the Alexander polynomial of the knots $N(T)$ and $D(T)$. In [17] an alternate definition is given that uses the bracket polynomial of the knots $N(T)$ and $D(T)$, and in [16] the fraction of a tangle is related to the conductance of an associated electrical network. In all these definitions the fraction is by definition an isotopy invariant of tangles. Below we discuss the bracket polynomial and coloring definitions of the fraction.

4.1 F(T) Through the Bracket Polynomial

In this section we shall discuss the structure of the the bracket state model for the Jones polynomial [20, 21] and how to construct the tangle fraction by using this technique. We first construct the bracket polynomial (state summation), which is a regular isotopy invariant (invariance under all but the Reidemeister move I). The bracket polynomial can be normalized to produce an invariant of all the Reidemeister moves. This invariant is known as the Jones polynomial [18, 19]. The Jones polynomial was originally discovered by a different method.

The *bracket polynomial*, $< K >=< K > (A)$, assigns to each unoriented link diagram K a Laurent polynomial in the variable A, such that

1. If K and K' are regularly isotopic diagrams, then $< K >=< K' >$.

2. If $K \amalg O$ denotes the disjoint union of K with an extra unknotted and unlinked component O (also called 'loop' or 'simple closed curve' or 'Jordan curve'), then

$$< K \amalg O >= \delta < K >,$$

where

$$\delta = -A^2 - A^{-2}.$$

3. $< K >$ satisfies the following formulas

$$< \chi >= A < \asymp > +A^{-1} <)(>$$
$$< \overline{\chi} >= A^{-1} < \asymp > +A <)(>,$$

where the small diagrams represent parts of larger diagrams that are identical except at the site indicated in the bracket. We take the convention that the letter chi, χ, denotes a crossing where *the curved line is crossing over the straight segment*. The barred letter denotes the switch of this crossing, where *the curved line is undercrossing the straight segment*. The above formulas can be summarized by the single equation

$$< K >= A < S_L K > +A^{-1} < S_R K > .$$

In this text formula we have used the notations $S_L K$ and $S_R K$ to indicate the two new diagrams created by the two smoothings of a single crossing in the diagram K. That is, K, $S_L K$ and $S_R K$ differ at the site of one crossing in the diagram K. These smoothings are described as follows. Label the four regions locally incident to a crossing by the letters L and R, with L labelling the region to the left of the undercrossing arc for a traveller who approaches the overcrossing on a route along the undercrossing arc. There are two such routes, one on each side of the overcrossing line. This labels two regions with L. The remaining two are labelled R. A smoothing is of *type L* if it connects the regions labelled L, and it is of *type R* if it connects the regions labelled R, see Figure 12.

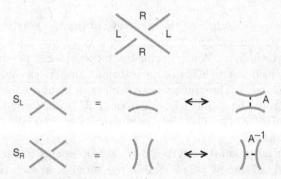

Fig. 12 Bracket smoothings.

It is easy to see that Properties 2 and 3 define the calculation of the bracket on arbitrary link diagrams. The choices of coefficients (A and A^{-1}) and the value of δ make the bracket invariant under the Reidemeister moves II and III (see [20]). Thus Property 1 is a consequence of the other two properties.

In order to obtain a closed formula for the bracket, we now describe it as a state summation. Let K be any unoriented link diagram. Define a *state*, S, of K to be a choice of smoothing for each crossing of K. There are two choices for smoothing a given crossing, and thus there are 2^N states of a diagram with N crossings. In a state we label each smoothing with A or A^{-1} according to the left-right convention discussed in Property 3 (see Figure 12). The label is called a *vertex weight* of the state. There are two evaluations related to a state. The first one is the product of the vertex weights, denoted

$$< K|S >.$$

The second evaluation is the number of loops in the state S, denoted

$$||S||.$$

Define the *state summation*, $< K >$, by the formula

$$< K > = \sum_S < K|S > \delta^{||S||-1}.$$

It follows from this definition that $< K >$ satisfies the equations

$$< \chi > = A < \asymp > + A^{-1} <)(>,$$

$$< K \amalg O > = \delta < K >,$$

$$< O > = 1.$$

The first equation expresses the fact that the entire set of states of a given diagram is the union, with respect to a given crossing, of those states with an A-type smoothing and those with an A^{-1}-type smoothing at that crossing. The second and the third equation are clear from the formula defining the state summation. Hence this state summation produces the bracket polynomial as we have described it at the beginning of the section.

In computing the bracket, one finds the following behaviour under Reidemeister move I:

$$< \gamma > = -A^3 < \smile >$$

and

$$< \overline{\gamma} > = -A^{-3} < \smile >$$

where γ denotes a curl of positive type as indicated in Figure 13, and $\overline{\gamma}$ indicates a curl of negative type, as also seen in this figure. The type of a curl is the sign of the crossing when we orient it locally. Our convention of signs is also given in Figure 13. Note that the type of a curl does not depend

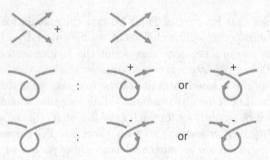

Fig. 13 Crossing, signs and curls.

on the orientation we choose. The small arcs on the right hand side of these formulas indicate the removal of the curl from the corresponding diagram.

The bracket is invariant under regular isotopy and can be normalized to an invariant of ambient isotopy by the definition

$$f_K(A) = (-A^3)^{-w(K)} < K > (A),$$

where we chose an orientation for K, and where $w(K)$ is the sum of the crossing signs of the oriented link K. $w(K)$ is called the *writhe* of K. The convention for crossing signs is shown in Figure 13.

By a change of variables one obtains the original Jones polynomial, $V_K(t)$, for oriented knots and links from the normalized bracket:

$$V_K(t) = f_K(t^{-\frac{1}{4}}).$$

The bracket model for the Jones polynomial is quite useful both theoretically and in terms of practical computations. One of the neatest applications is to simply compute $f_K(A)$ for the trefoil knot T and determine that $f_T(A)$ is not equal to $f_T(A^{-1}) = f_{-T}(A)$. This shows that the trefoil is not ambient isotopic to its mirror image, a fact that is quite tricky to prove by classical methods.

For 2-tangles, we do smoothings on the tangle diagram until there are no crossings left. As a result, *a state of a 2-tangle* consists in a collection of loops in the tangle box, plus simple arcs that connect the tangle ends. The loops evaluate to powers of δ, and what is left is either the tangle [0] or the tangle [∞], since [0] and [∞] are the only ways to connect the tangle inputs and outputs without introducing any crossings in the diagram. In analogy to knots and links, we can find a *state summation* formula for the *bracket of the tangle*, denoted $< T >$, by summing over the states obtained by smoothing each crossing in the tangle. For this we define the *remainder of a state*, denoted R_S, to be either the tangle [0] or the tangle [∞]. Then the evaluation of $< T >$ is given by

$$< T > = \sum_S < T|S > \delta^{||S||} < R_S >,$$

where $< T|S >$ is the product of the vertex weights (A or A^{-1}) of the state S of T. The above formula is consistent with the formula for knots obtained by taking the closure $N(T)$ or $D(T)$. In fact, we have the following formula:

$$< N(T) > = \sum_S < T|S > \delta^{||S||} < N(R_S) > .$$

Note that $< N([0]) > = \delta$ and $< N([\infty]) > = 1$. A similar formula holds for $< D(T) >$. Thus, $< T >$ appears as a linear combination with Laurent polynomial coefficients of $< [0] >$ and $< [\infty] >$, i.e. $< T >$ takes values in the free module over $\mathbb{Z}[A, A^{-1}]$ with basis $\{< [0] >, < [\infty] >\}$. Notice that two elements in this module are equal iff the corresponding coefficients of the basis elements coincide. Note also that $< T >$ is an invariant of regular isotopy with values in this module. We have just proved the following:

Lemma 4.1. *Let T be any 2-tangle and let $< T >$ be the formal expansion of the bracket on this tangle. Then there exist elements $n_T(A)$ and $d_T(A)$ in $\mathbb{Z}[A, A^{-1}]$, such that*

$$< T > = d_T(A) < [0] > + n_T(A) < [\infty] >,$$

and $n_T(A)$ and $d_T(A)$ are regular isotopy invariants of the tangle T.

In order to evaluate $< N(T) >$ in the formula above we need only apply the closure N to $[0]$ and $[\infty]$. More precisely, we have:

Lemma 4.2. $< N(T) > = d_T \delta + n_T$ *and* $< D(T) > = d_T + n_T \delta$.

Proof. Since the smoothings of crossings do not interfere with the closure (N or D), the closure will carry through linearly to the whole sum of $< T >$. Thus,

$$< N(T) > = d_T(A) < N([0]) > + n_T(A) < N([\infty]) > = d_T(A)\delta + n_T(A),$$
$$< D(T) > = d_T(A) < D([0]) > + n_T(A) < D([\infty]) > = d_T(A) + n_T(A)\delta. \qquad \square$$

We define now the *polynomial fraction, $frac_T(A)$*, of the 2-tangle T to be the ratio

$$frac_T(A) = \frac{n_T(A)}{d_T(A)}$$

in the ring of fractions of $\mathbb{Z}[A, A^{-1}]$ with a formal symbol ∞ adjoined.

Lemma 4.3. $frac_T(A)$ *is an invariant of ambient isotopy for 2-tangles.*

Proof. Since d_T and n_T are regular isotopy invariants of T, it follows that $frac_T(A)$ is also a regular isotopy invariant of T. Suppose now $T\gamma$ is T with a curl added. Then $< T\gamma > = (-A^3) < T >$ (same remark for $\bar{\gamma}$). So, $n_{T\gamma}(A) = -A^3 n_T(A)$ and $d_{T\gamma}(A) = -A^3 d_T(A)$. Thus, $n_{T\gamma}/d_{T\gamma} = n_T/d_T$. This shows that $frac_T$ is also invariant under the Reidemeister move I, and hence an ambient isotopy invariant. $\qquad \square$

Lemma 4.4. *Let T and S be two 2-tangles. Then, we have the following formula for the bracket of the sum of the tangles.*

$$< T + S > = d_T d_S < [0] > + (d_T n_S + n_T d_S + n_S \delta) < [\infty] > .$$

Thus

$$frac_{T+S} = frac_T + frac_S + \frac{n_S \delta}{d_T d_S}.$$

Proof. We do first the smoothings in T leaving S intact, and then in S:

$$< T + S > = d_T < [0] + S > + n_T < [\infty] + S >$$

$$= d_T < S > + n_T < [\infty] + S >$$

$$= d_T (d_S < [0] > + n_S < [\infty] >)$$

$$+ n_T (d_S < [\infty] + [0] > + n_S < [\infty] + [\infty] >)$$

$$= d_T (d_S < [0] > + n_S < [\infty] >) + n_T (d_S < [\infty] > + n_S \delta < [\infty] >)$$

$$= d_T d_S < [0] > + (d_T n_S + n_T d_S + n_S \delta) < [\infty] > .$$

Thus, $n_{T+S} = (d_T n_S + n_T d_S + n_S \delta)$ and $d_{T+S} = d_T d_S$. A straightforward calculation gives now $frac_{T+S}$. □

As we see from Lemma 4, $frac_T(A)$ will be additive on tangles if

$$\delta = -A^2 - A^{-2} = 0.$$

Moreover, from Lemma 2 we have for $\delta = 0$, $< N(T) > = n_T$, $< D(T) > = d_T$. This nice situation will be our main object of study in the rest of this section. Now, if we set $A = \sqrt{i}$ where $i^2 = -1$, then it is

$$\delta = -A^2 - A^{-2} = -i - i^{-1} = -i + i = 0.$$

For this reason, we shall henceforth assume that A takes the value $\sqrt{i}$. So $< K >$ will denote $< K > (\sqrt{i})$ for any knot or link K.

We now define the 2-*tangle fraction* $F(T)$ by the following formula:

$$nF(T) = i \frac{n_T(\sqrt{i})}{d_T(\sqrt{i})}.$$

We will let $n(T) = n_T(\sqrt{i})$ and $d(T) = d_T(\sqrt{i})$, so that

$$F(T) = i \frac{n(T)}{d(T)}.$$

Lemma 4.5. *The 2-tangle fraction has the following properties.*

1. $F(T) = i < N(T) > / < D(T) >$, *and it is a real number or* ∞,
2. $F(T + S) = F(T) + F(S)$,
3. $F([0]) = \frac{0}{1}$,

4. $F([1]) = \frac{1}{1}$,
5. $F([\infty]) = \frac{1}{0}$,
6. $F(-T) = -F(T)$, *in particular* $F([-1]) = -\frac{1}{1}$,
7. $F(1/T) = 1/F(T)$,
8. $F(T^r) = -1/F(T)$.

As a result we conclude that for a tangle obtained by arithmetic operations from integer tangles $[n]$, the fraction of that tangle is the same as the fraction obtained by doing the same operations to the corresponding integers. (This will be studied in detail in the next section.)

Proof. The formula $F(T) = i < N(T) > / < D(T) >$ and Statement 2. follow from the observations above about $\delta = 0$. In order to show that $F(T)$ is a real number or ∞ we first consider $< K >:=< K > (\sqrt{i})$, for K a knot or link, as in the hypotheses prior to the lemma. Then we apply this information to the ratio $i < N(T) > / < D(T) >$.

Let K be any knot or link. We claim that then $< K >= \omega p$, where ω is a power of $\sqrt{i}$ and p is an integer. In fact, we will show that each non-trivial state of K contributes $\pm \omega$ to $< K >$. In order to show this, we examine how to get from one non-trivial state to another. It is a fact that, for any two states, we can get from one to the other by resmoothing a subset of crossings. It is possible to get from any single loop state (and only single loop states of K contribute to $< K >$, since $\delta = 0$) to any other single loop state by a series of *double resmoothings*. In a double resmoothing we resmooth two crossings, such that one of the resmoothings disconnects the state and the other reconnects it. See Figure 14 for an illustration. Now consider the effect of a double resmoothing on the evaluation of one state. Two crossings change. If one is labelled A and the other A^{-1}, then there is no net change in the evaluation of the state. If both are A, then we go from $A^2 P$ (P is the rest of the product of state labels) to $A^{-2}P$. But $A^2 = i$ and $A^{-2} = -i$. Thus if one state contributes $\omega = ip$, then the other state contributes $-\omega = -ip$. These remarks prove the claim.

Now, a state that contributes non-trivially to $N(T)$ must have the form of the tangle $[\infty]$. We will show that if S is a state of T contributing non-trivially to $< N(T) >$ and S' a state of T contributing non-trivially to $< D(T) >$, then $< S > / < S' > = \pm i$. Here $< S >$ denotes the product of the vertex weights for S, and $< S' >$ is the product of the vertex weights for S'.

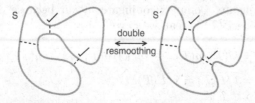

Fig. 14 A double resmoothing.

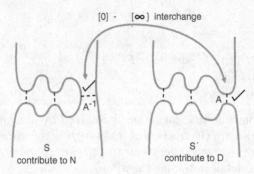

Fig. 15 Non-trivial states.

If this ratio is verified for some pair of states S, S', then it follows from the first claim that it is true for all pairs of states, and that $< N(T) >= \omega p$, $< D(T) >= \omega' q$, $p, q \in \mathbb{Z}$ and $\omega/\omega' = < S >/< S' > = \pm i$. Hence $< N(T) >/< D(T) > = \pm i\, p/q$, where p/q is a rational number (or $q = 0$). This will complete the proof that $F(T)$ is real or ∞.

To see this second claim we consider specific pairs of states as in Figure 15. We have illustrated representative states S and S' of the tangle T. We obtain S' from S by resmoothing at one site that changes S from an $[\infty]$ tangle to the $[0]$ tangle underlying S'. Then $< S >/< S' > = A^{\pm 2} = \pm i$. If there is no such resmoothing site available, then it follows that $D(T)$ is a disjoint union of two diagrams, and hence $< D(T) > = 0$ and $F(T) = \infty$. This does complete the proof of Statement 1.

At $\delta = 0$ we also have:

$$< N([0] >= 0, \ < D([0]) >= 1, \ < N([\infty]) >= 1, \ < D([\infty]) >= 0,$$ and so, the evaluations 3. to 5. are easy. For example, note that

$$< [1] > = A < [0] > + A^{-1} < [\infty] >,$$

hence

$$F([1]) = i\,\frac{A^{-1}}{A} = i\,A^{-2} = i\,(i^{-1}) = 1.$$

To have the fraction value 1 for the tangle $[1]$ is the reason that in the definition of $F(T)$ we normalized by i. Statement 6. follows from the fact that the bracket of the mirror image of a knot K is the same as the bracket of K, but with A and A^{-1} switched. For proving 7. we observe first that for any 2-tangle T, $d(\frac{1}{T}) = \overline{n(T)}$ and $n(\frac{1}{T}) = \overline{d(T)}$, where the overline denotes the complex conjugate. Complex conjugates occur because $A^{-1} = \overline{A}$ when $A = \sqrt{i}$. Now, since $F(T)$ is real, we have

$$F(\tfrac{1}{T}) = i\overline{d(T)}/\overline{n(T)} = \overline{-id(T)/n(T)} = \overline{1/(i\,n(T)/d(T))}$$
$$= \overline{1/F(T)} = 1/F(T).$$

Statement 8. follows immediately from 6. and 7. This completes the proof.

$\square$

For a related approach to the well-definedness of the 2-tangle fraction, the reader should consult [31]. The double resmoothing idea originates from [24].

Remark 4.6. For any knot or link K we define the *determinant* of K by the formula

$$Det(K) := | < K > (\sqrt{i})|$$

where $|z|$ denotes the modulus of the complex number z. Thus we have the formula

$$|F(T)| = \frac{Det(N(T))}{Det(D(T))}$$

for any 2-tangle T.

In other approaches to the theory of knots, the determinant of the knot is actually the determinant of a certain matrix associated either to the diagram for the knot or to a surface whose boundary is the knot. See [43, 23] for more information on these connections. Conway's original definition of the fraction [7] is $\Delta_{N(T)}(-1)/\Delta_{D(T)}(-1)$ where $\Delta_K(-1)$ denotes the evaluation of the Alexander polynomial of a knot K at the value -1. In fact, $|\Delta_K(-1)| = Det(K)$, and with appropriate attention to signs, the Conway definition and our definition using the bracket polynomial coincide for all 2-tangles.

4.2 The Fraction through Coloring

We conclude this section by giving an alternate definition of the fraction that uses the concept of coloring of knots and tangles. We color the arcs of the knot/tangle with integers, using the basic coloring rule that if two undercrossing arcs colored α and γ meet at an overcrossing arc colored β, then $\alpha + \gamma = 2\beta$. We often think of one of the undercrossing arc colors as determined by the other two colors. Then one writes $\gamma = 2\beta - \alpha$.

It is easy to verify that this coloring method is invariant under the Reidemeister moves in the following sense: Given a choice of coloring for the tangle/knot, there is a way to re-color it each time a Reidemeister move is performed, so that no change occurs to the colors on the external strands of the tangle (so that we still have a valid coloring). This means that a coloring potentially contains topological information about a knot or a tangle. In coloring a knot (and also many non-rational tangles) it is usually necessary to restrict the colors to the set of integers modulo N for some modulus N. For example, in Figure 16 it is clear that the color set $\mathbb{Z}/3\mathbb{Z} = \{0, 1, 2\}$ is forced for coloring a trefoil knot. When there exists a coloring of a tangle by integers, so that it is not necessary to reduce the colors over some modulus we shall say that the tangle is *integrally colorable*.

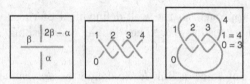

Fig. 16 The coloring rule, integral and modular coloring.

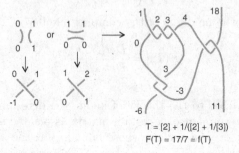

$$T = [2] + 1/([2] + 1/[3])$$
$$F(T) = 17/7 = f(T)$$

Fig. 17 Coloring rational tangles.

It turns out that *every rational tangle is integrally colorable:* To see this choose two 'colors' for the initial strands (e.g. the colors 0 and 1) and color the rational tangle as you create it by successive twisting. We call the colors on the initial strands the *starting colors.* See Figure 17 for an example. It is important that we start coloring from the initial strands, because then the coloring propagates automatically and uniquely. If one starts from somewhere else, one might get into an edge with an undetermined color. The resulting colored tangle now has colors assigned to its external strands at the northwest, northeast, southwest and southeast positions. Let $NW(T)$, $NE(T)$, $SW(T)$ and $SE(T)$ denote these respective colors of the colored tangle T and define the *color matrix of T, $M(T)$*, by the equation

$$M(T) = \begin{bmatrix} NW(T) \ NE(T) \\ SW(T) \ SE(T) \end{bmatrix}.$$

Definition 4.7. To a rational tangle T with color matrix $M(T) = \begin{bmatrix} a \ b \\ c \ d \end{bmatrix}$ we associate the number

$$f(T) := \frac{b-a}{b-d} \in \mathbb{Q} \cup \infty.$$

It turns out that the entries a, b, c, d of a color matrix of a rational tangle satisfy the 'diagonal sum rule': $a + d = b + c$.

Proposition 4.8. *The number $f(T)$ is a topological invariant associated with the tangle T. In fact, $f(T)$ has the following properties:*

1. $f(T + [\pm 1]) = f(T) \pm 1$,

2. $f(-\frac{1}{T}) = -\frac{1}{f(T)}$,

3. $f(-T) = -f(T)$,

4. $f(\frac{1}{T}) = \frac{1}{f(T)}$,

5. $f(T) = F(T)$.

Thus the coloring fraction is identical to the arithmetical fraction defined earlier.

It is easy to see that $f([0]) = \frac{0}{1}$, $f([\infty]) = \frac{1}{0}$, $f([\pm 1]) = \pm 1$. Hence Statement 5 follows by induction. For proofs of all statements above as well as for a more general set-up we refer the reader to our paper [26]. This definition is quite elementary, but applies only to rational tangles and tangles generated from them by the algebraic operations of '$+$' and '$*$'.

In Figure 17 we have illustrated a coloring over the integers for the tangle $[[2], [2], [3]]$ such that every edge is labelled by a different integer. This is always the case for an alternating rational tangle diagram T. For the numerator closure $N(T)$ one obtains a coloring in a modular number system. For example in Figure 17 the coloring of $N(T)$ will be in $\mathbb{Z}/17\mathbb{Z}$, and it is easy to check that the labels remain distinct in this example. For rational tangles, this is always the case when $N(T)$ has a prime determinant, see [26] and [37]. It is part of a more general conjecture about alternating knots and links [25, 1].

4.3 The Fraction through Conductance

Conductance is a quantity defined in electrical networks as the inverse of resistance. For pure resistances, conductance is a positive quantity. Negative conductance corresponds to amplification, and is commonly included in the physical formalism. One defines the conductance between two vertices in a graph (with positive or negative conductance weights on the edges of the graph) as a sum of weighted trees in the graph divided by a sum of weighted trees of the same graph, but with the two vertices identified. This definition allows negative values for conductance and it agrees with the classical one. Conductance satisfies familiar laws of parallel and series connection as well as a star-triangle relation.

By associating to a given knot or link diagram the corresponding signed checkerboard graph (see [26, 16] for a definition of this well-known association of graph to link diagram), one can define [16] the conductance of a knot or link between any two regions that receive the same color in the checkerboard graph. The conductance of the link between these two regions is an isotopy invariant of the link (with motion restricted to Reidemeister moves that do not pass across the selected regions). This invariance follows from properties of series/parallel connection and the star-triangle relation. These circuit laws turn out to be images of the Reidemeister moves under the translation from knot or link diagram to checkerboard graph! For a 2-tangle we take the conductance to be the conductance of the numerator of the tangle, between the two bounded regions adjacent to the closures at the top and bottom of the tangle.

The conductance of a 2-tangle turns out to be the same as the fraction of the tangle. This provides yet another way to define and verify the isotopy invariance of the tangle fraction for any 2-tangle.

5 The Classification of Unoriented Rational Knots

By taking their numerators or denominators rational tangles give rise to a special class of knots, the rational knots. We have seen so far that rational tangles are directly related to finite continued fractions. We carry this insight further into the classification of rational knots (Schubert's theorems). In this section we consider unoriented knots, and by Remark 3.1 we will be using the 3-strand-braid representation for rational tangles with odd number of terms. Also, by Lemma 2, we may assume all rational knots to be alternating. Note that we only need to take numerator closures, since the denominator closure of a tangle is simply the numerator closure of its rotate.

As already said in the introduction, it may happen that two rational tangles are non-isotopic but have isotopic numerators. The simplest instance of this phenomenon is adding n twists at the bottom of a tangle T, see Figure 18. This operation does not change the knot $N(T)$, i.e. $N(T * 1/[n]) \sim N(T)$, but it does change the tangle, since $F(T * 1/[n]) = F(1/([n] + 1/T)) = 1/(n + 1/F(T))$; so, if $F(T) = p/q$, then $F(T * 1/[n]) = p/(np + q)$. Hence, if we set $np + q = q'$ we have $q \equiv q'(mod\, p)$, just as Theorem 2 dictates. Note that reducing all possible bottom twists implies $|p| > |q|$.

Another key example of the arithmetic relationship of the classification of rational knots is illustrated in Figure 19. Here we see that the 'palindromic' tangles

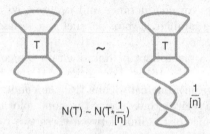

$$N(T) \sim N(T * \frac{1}{[n]})$$

Fig. 18 Twisting the bottom of a tangle.

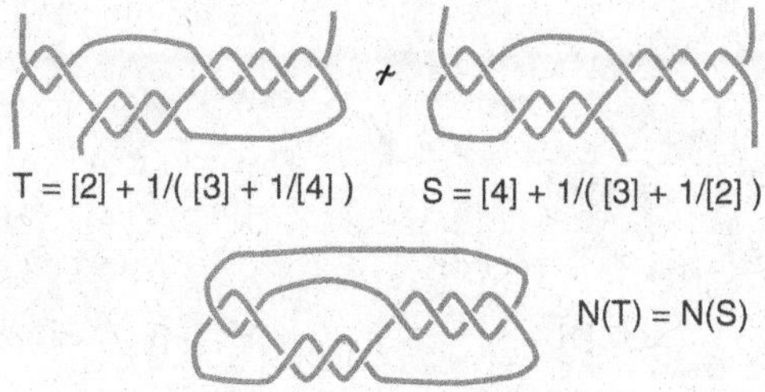

$$T = [2] + 1/(\,[3] + 1/[4]\,) \qquad S = [4] + 1/(\,[3] + 1/[2]\,)$$

$$N(T) = N(S)$$

Fig. 19 An instance of the palindrome equivalence.

$$T = [[2], [3], [4]] = [2] + \cfrac{1}{[3] + \frac{1}{[4]}}$$

and

$$S = [[4], [3], [2]] = [4] + \cfrac{1}{[3] + \frac{1}{[2]}}$$

both close to the same rational knot, shown at the bottom of the figure. The two tangles are different, since they have different corresponding fractions:

$$F(T) = 2 + \cfrac{1}{3 + \frac{1}{4}} = \frac{30}{13} \quad \text{and} \quad F(S) = 4 + \cfrac{1}{3 + \frac{1}{2}} = \frac{30}{7}.$$

Note that the product of 7 and 13 is congruent to 1 modulo 30.
More generally, consider the following two fractions:

$$F = [a, b, c] = a + \cfrac{1}{b + \frac{1}{c}} \quad \text{and} \quad G = [c, b, a] = c + \cfrac{1}{b + \frac{1}{a}}.$$

We find that

$$F = a + c\,\frac{1}{cb + 1} = \frac{abc + a + c}{bc + 1} = \frac{P}{Q},$$

while

$$G = c + a\,\frac{1}{ab + 1} = \frac{abc + c + a}{ab + 1} = \frac{P}{Q'}.$$

Thus we found that $F = \frac{P}{Q}$ and $G = \frac{P}{Q'}$, where

$$QQ' = (bc + 1)(ab + 1) = ab^2c + ab + bc + 1 = bP + 1.$$

Assuming that a, b and c are integers, we conclude that

$$QQ' \equiv 1\,(mod\,P).$$

This pattern generalizes to arbitrary continued fractions and their palindromes (obtained by reversing the order of the terms). I.e. If $\{a_1, a_2, \ldots, a_n\}$ *is a collection of n non-zero integers, and if $A = [a_1, a_2, \ldots, a_n] = \frac{P}{Q}$ and*

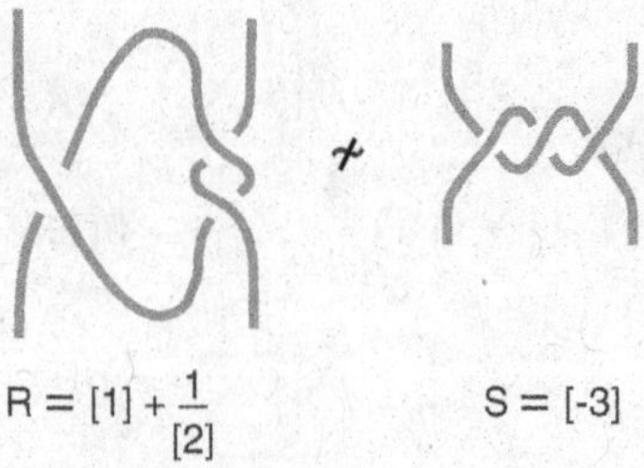

$$R = [1] + \frac{1}{[2]} \qquad\qquad S = [-3]$$

Fig. 20 An example of the special cut.

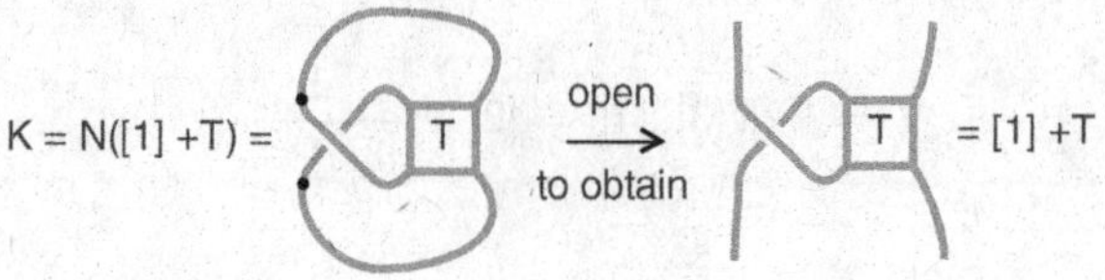

Fig. 21 A standard cut.

$B = [a_n, a_{n-1}, \ldots, a_1] = \frac{P'}{Q'}$, then $P = P'$ and $QQ' \equiv (-1)^{n+1} (mod\, P)$. We will be referring to this as 'the Palindrome Theorem'. The Palindrome Theorem is a known result about continued fractions. For example, see [47] and [27]. Note that we need n to be odd in the previous congruence. This agrees with Remark 3.1 that without loss of generality the terms in the continued fraction of a rational tangle may be assumed to be odd.

Finally, Figure 20 illustrates another basic example for the unoriented Schubert Theorem. The two tangles $R = [1] + \frac{1}{[2]}$ and $S = [-3]$ are non-isotopic by the Conway Theorem, since $F(R) = 1 + 1/2 = 3/2$ while $F(S) = -3 = 3/ - 1$. But they have isotopic numerators: $N(R) \sim N(S)$, the left-handed trefoil. Now 2 is congruent to -1 modulo 3, confirming Theorem 2.

We now analyse the above example in general. From the analysis of the bottom twists we can assume without loss of generality that a rational tangle R has fraction $\frac{P}{Q}$, for $|P| > |Q|$. Thus R can be written in the form $R = [1] + T$ or $R = [-1] + T$. We consider the rational knot diagram $K = N([1] + T)$, see Figure 21. (We analyze $N([-1] + T)$ in the same way.) The tangle $[1] + T$ is said to arise as a *standard cut* on K.

Notice that the indicated horizontal crossing of $N([1] + T)$ could be also seen as a vertical one. So, we could also cut the diagram K at the two other marked points (see Figure 22) and still obtain a rational tangle, since T is rational. The tangle obtained by cutting K in this second pair of points is said to arise as a *special cut* on K. Figure 22 demonstrates that the tangle of the special cut is the tangle $[-1] - 1/T$. So we have $N([1] + T) \sim N([-1] - \frac{1}{T})$. Suppose now $F(T) = p/q$. Then $F([1] + T) = 1 + p/q = (p + q)/q$, while $F([-1] - 1/T) = -1 - q/p = (p + q)/(-p)$, so the two rational tangles that give rise to the same knot K are not isotopic. Since $-p \equiv q \, mod(p + q)$, this

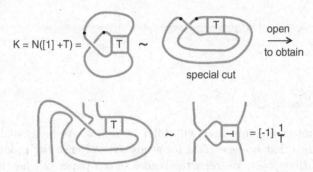

Fig. 22 A special cut.

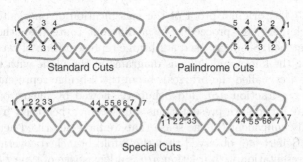

Standard Cuts   Palindrome Cuts

Special Cuts

Fig. 23 Standard, palindrome and special cuts.

equivalence is another example for Theorem 2. In Figure 22 if we took $T = \frac{1}{[2]}$ then $[-1] - 1/T = [-3]$ and we would obtain the example of Figure 20.

The proof of Theorem 2 can now proceed in two stages. First, given a rational knot diagram we look for all possible places where we could cut and open it to a rational tangle. The crux of our proof in [27] is the fact that all possible 'rational cuts' on a rational knot fall into one of the basic cases that we have already discussed. I.e. we have the standard cuts, the palindrome cuts and the special cuts. In Figure 23 we illustrate on a representative rational knot, all the cuts that exhibit that knot as a closure of a rational tangle. Each pair of points is marked with the same number. The arithmetics is similar to the cases that have been already verified. It is convenient to say that reduced fractions p/q and p'/q' are *arithmetically equivalent*, written $p/q \sim p'/q'$ if $p = p'$ and either $qq' \equiv 1 \pmod{p}$ or $q \equiv q' \pmod{p}$. In this language, Schubert's theorem states that two rational tangles close to form isotopic knots if and only if their fractions are arithmetically equivalent.

In Figure 24 we illustrate one example of a cut that is not allowed since it opens the knot to a non-rational tangle.

In the second stage of the proof we want to check the arithmetic equivalence for two different given knot diagrams, numerators of some rational tangles. By Lemma 2 the two knot diagrams may be assumed alternating,

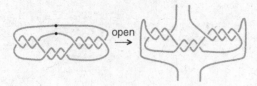

Fig. 24 A non-rational cut.

so by the Tait Conjecture they will differ by flypes. We analyse all possible flypes to prove that no new cases for study arise. Hence the proof becomes complete at that point. We refer the reader to our paper [27] for the details.

$\square$

Remark 5.1. The original proof of the classification of unoriented rational knots by Schubert [46] proceeded by a different route than the proof we have just sketched. Schubert used a 2-bridge representation of rational knots (representing the knots and links as diagrams in the plane with two special overcrossing arcs, called the bridges). From the 2-bridge representation, one could extract a fraction p/q, and Schubert showed by means of a canonical form, that if two such presentations are isotopic, then their fractions are arithmetically equivalent (in the sense that we have described here). On the other hand, Seifert [46] observed that the 2-fold branched covering space of a 2-bridge presentation with fraction p/q is a lens space of type $L(p, q)$. Lens spaces are a particularly tractable set of three manifolds, and it is known by work of Reidemeister and Franz [42, 15] that $L(p, q)$ is homeomorphic to $L(p', q')$ if and only if p/q and p'/q' are arithmetically equivalent. Furthermore, one knows that if knots K and K' are isotopic, then their 2-fold branched covering spaces are homeomorphic. Hence it follows that if two rational knots are isotopic, then their fractions are arithmetically equivalent (via the result of Reidemeister and Franz classifying lens spaces). In this way Schubert proved that two rational knots are isotopic if and only if their fractions are arithmetically equivalent.

6 Rational Knots and Their Mirror Images

In this section we give an application of Theorem 2. An unoriented knot or link K is said to be *achiral* if it is topologically equivalent to its mirror image $-K$. If a link is not equivalent to its mirror image then it is said be *chiral*. One then can speak of the *chirality* of a given knot or link, meaning whether it is chiral or achiral. Chirality plays an important role in the applications of Knot Theory to Chemistry and Molecular Biology. It is interesting to use the classification of rational knots and links to determine their chirality. Indeed, we have the following well-known result (for example see [47] and also page 24, Exercise 2.1.4 in [28]):

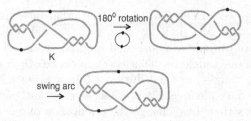

Fig. 25 An achiral rational link.

Theorem 6.1. *Let $K = N(T)$ be an unoriented rational knot or link, presented as the numerator of a rational tangle T. Suppose that $F(T) = p/q$ with p and q relatively prime. Then K is achiral if and only if $q^2 \equiv -1 \,(mod\,p)$. It follows that achiral rational knots and links are all numerators of rational tangles of the form $[[a_1], [a_2], \ldots, [a_k], [a_k], \ldots, [a_2], [a_1]]$ for any integers $a_1, \ldots, a_k$.*

Note that in this description we are using a representation of the tangle with an even number of terms. The leftmost twists $[a_1]$ are horizontal, thus the rightmost starting twists $[a_1]$ are vertical.

Proof. With $-T$ the mirror image of the tangle T, we have that $-K = N(-T)$ and $F(-T) = p/(-q)$. If K is topologically equivalent to $-K$, then $N(T)$ and $N(-T)$ are equivalent, and it follows from the classification theorem for rational knots that either $q(-q) \equiv 1 \,(mod\,p)$ or $q \equiv -q \,(mod\,p)$. Without loss of generality we can assume that $0 < q < p$. Hence $2q$ is not divisible by p and therefore it is not the case that $q \equiv -q \,(mod\,p)$. Hence $q^2 \equiv -1 \,(mod\,p)$. Conversely, if $q^2 \equiv -1 \,(mod\,p)$, then it follows from the Palindrome Theorem (described in the previous section) [27] that *the continued fraction expansion of p/q has to be symmetric with an even number of terms*. It is then easy to see that the corresponding rational knot or link, say $K = N(T)$, is equivalent to its mirror image. One rotates K by $180°$ in the plane and swings an arc, as Figure 25 illustrates. This completes the proof. $\square$

In [12] the authors find an explicit formula for the number of achiral rational knots among all rational knots with n crossings.

7 The Oriented Case

Oriented rational knots and links arise as numerator closures of oriented rational tangles. In order to compare oriented rational knots via rational tangles we need to examine how rational tangles can be oriented. We orient rational tangles by choosing an orientation for each strand of the tangle. Here we are only interested in orientations that yield consistently oriented knots upon taking the numerator closure. This means that the two top end

arcs have to be oriented one inward and the other outward. Same for the
two bottom end arcs. We shall say that two oriented rational tangles are
isotopic if they are isotopic as unoriented tangles, by an isotopy that carries
the orientation of one tangle to the orientation of the other. Note that, since
the end arcs of a tangle are fixed during a tangle isotopy, this means that
the tangles must have identical orientations at their four end arcs *NW, NE,
SW, SE*. It follows that if we change the orientation of one or both strands
of an oriented rational tangle we will always obtain a non-isotopic oriented
rational tangle.

Reversing the orientation of one strand of an oriented rational tangle may
or may not give rise to isotopic oriented rational knots. Figure 26 illustrates
an example of non-isotopic oriented rational knots, which are isotopic as
unoriented knots.

Reversing the orientation of both strands of an oriented rational tangle will
always give rise to two isotopic oriented rational knots or links. We can see
this by doing a vertical flip, as Figure 27 demonstrates. Using this observation
we conclude that, as far as the study of oriented rational knots is concerned,
*all oriented rational tangles may be assumed to have the same orientation
for their NW and NE end arcs.* We fix this orientation to be downward for
the *NW* end arc and upward for the *NE* arc, as in the examples of Figure 26
and as illustrated in Figure 28. Indeed, if the orientations are opposite of the
fixed ones doing a vertical flip the knot may be considered as the numerator
of the vertical flip of the original tangle. But this is unoriented isotopic to
the original tangle (recall Section 2, Figure 7), whilst its orientation pattern
agrees with our convention.

Thus we reduce our analysis to two basic types of orientation for the four
end arcs of a rational tangle. We shall call an oriented rational tangle of *type*

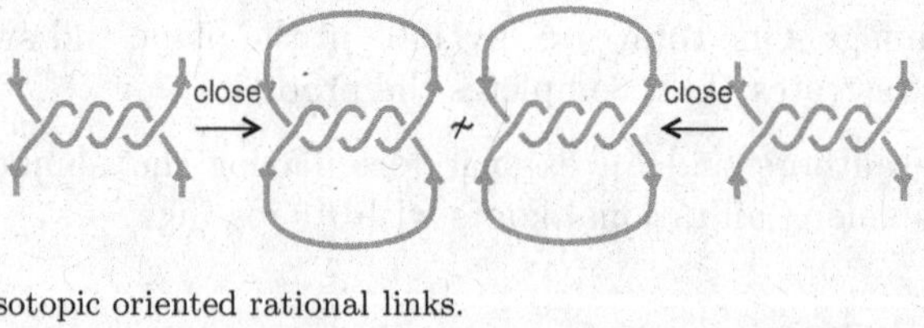

Fig. 26 Non-isotopic oriented rational links.

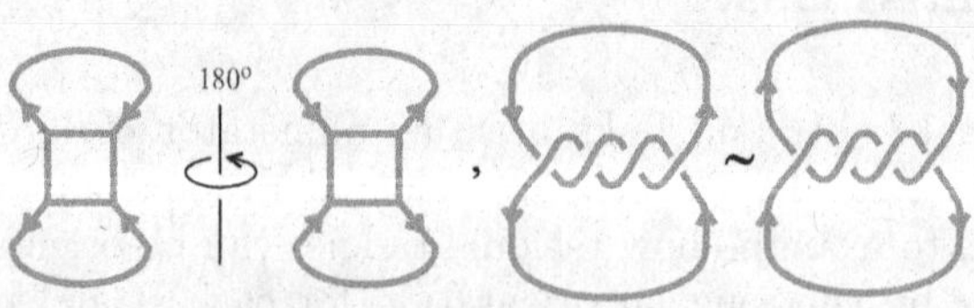

Fig. 27 Isotopic oriented rational knots and links.

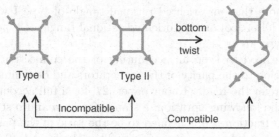

Fig. 28 Compatible and incompatible orientations.

I if the *SW* arc is oriented upward and the *SE* arc is oriented downward, and of *type II* if the *SW* arc is oriented downward and the *SE* arc is oriented upward, see Figure 28. From the above remarks, any tangle is of type I or type II. Two tangles are said to be *compatible* it they are both of type I or both of type II and *incompatible* if they are of different types. In order to classify oriented rational knots seen as numerator closures of oriented rational tangles, we will always compare compatible rational rangles. Note that if two oriented tangles are incompatible, adding a single half twist at the bottom of one of them yields a new pair of compatible tangles, as Figure 28 illustrates. Note also that adding such a twist, although it changes the tangle, it does not change the isotopy type of the numerator closure. Thus, up to bottom twists, we are always able to compare oriented rational tangles of the same orientation type.

We shall now introduce the notion of *connectivity* and we shall relate it to orientation and the fraction of unoriented rational tangles. We shall say that an unoriented rational tangle has *connectivity* type [0] if the *NW* end arc is connected to the *NE* end arc and the *SW* end arc is connected to the *SE* end arc. Similarly, we say that the tangle has *connectivity* type [+1] or type [∞] if the end arc connections are the same as in the tangles [+1] and [∞] respectively. The basic connectivity patterns of rational tangles are exemplified by the tangles [0], [∞] and [+1]. We can represent them iconically by the symbols shown below.

$$[0] = \asymp$$
$$[\infty] = ><$$
$$[+1] = \chi$$

Note that connectivity type [0] yields two-component rational links, while type [+1] or [∞] yields one-component rational links. Also, adding a bottom twist to a rational tangle of connectivity type [0] will not change the connectivity type of the tangle, while adding a bottom twist to a rational tangle of connectivity type [∞] will switch the connectivity type to [+1] and vice versa. While the connectivity type of unoriented rational tangles may be [0],

[+1] or [∞], note that an oriented rational tangle of type I will have connectivity type [0] or [∞] and an oriented rational tangle of type II will have connectivity type [0] or [+1].

Further, we need to keep an accounting of the connectivity of rational tangles in relation to the parity of the numerators and denominators of their fractions. We refer the reader to our paper [27] for a full account.

We adopt the following notation: e stands for *even* and o stands for *odd*. The *parity* of a fraction p/q is defined to be the ratio of the parities (e or o) of its numerator and denominator p and q. Thus the fraction $2/3$ is of parity e/o. The tangle [0] has fraction $0 = 0/1$, thus parity e/o, the tangle [∞] has fraction $\infty = 1/0$, thus parity o/e, and the tangle [+1] has fraction $1 = 1/1$, thus parity o/o. We then have the following result.

Theorem 7.1. *A rational tangle T has connectivity type $\asymp$ if and only if its fraction has parity e/o. T has connectivity type $><$ if and only if its fraction has parity o/e. T has connectivity type χ if and only if its fraction has parity o/o. (Note that the formal fraction of [∞] itself is $1/0$.) Thus the link $N(T)$ has two components if and only if T has fraction $F(T)$ of parity e/o.*

We will now proceed with sketching the proof of Theorem 3. We shall prove Schubert's oriented theorem by appealing to our previous work on the unoriented case and then analyzing how orientations and fractions are related. Our strategy is as follows: Consider an oriented rational knot or link diagram K in the form $N(T)$ where T is a rational tangle in continued fraction form. Then any other rational tangle that closes to this knot $N(T)$ is available, up to bottom twists if necessary, as a cut from the given diagram. If two rational tangles close to give K as an unoriented rational knot or link, then there are orientations on these tangles, induced from K so that the oriented tangles close to give K as an oriented knot or link. The two tangles may or may not be compatible. Thus, we must analyze when, comparing with the standard cut for the rational knot or link, another cut produces a compatible or incompatible rational tangle. However, assuming the top orientations are the same, we can replace one of the two incompatible tangles by the tangle obtained by adding a twist at the bottom. *It is this possible twist difference that gives rise to the change from modulus p in the unoriented case to the modulus $2p$ in the oriented case.* We will now perform this analysis. There are many interesting aspects to this analysis and we refer the reader to our paper [27] for these details. Schubert [46] proved his version of the oriented theorem by using the 2-bridge representation of rational knots and links, see also [6]. We give a tangle-theoretic combinatorial proof based upon the combinatorics of the unoriented case.

The simplest instance of the classification of oriented rational knots is adding an *even number of twists* at the bottom of an oriented rational tangle T, see Figure 28. We then obtain a compatible tangle $T * 1/[2n]$, and $N(T * 1/[2n]) \sim N(T)$. If now $F(T) = p/q$, then $F(T*1/[2n]) = F(1/([2n]+1/T)) = 1/(2n + 1/F(T)) = p/(2np + q)$. Hence, if we set $2np + q = q'$ we have

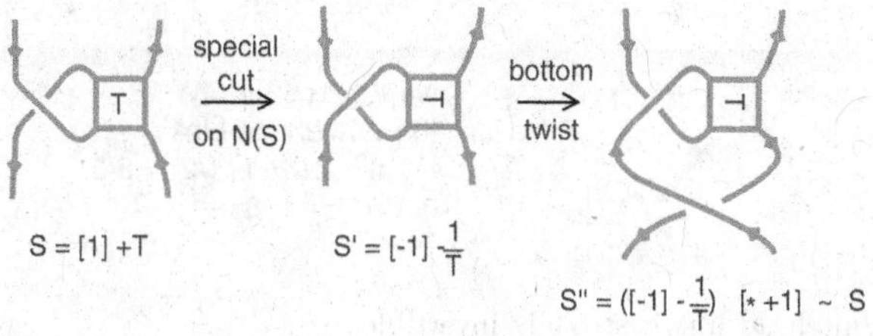

Fig. 29 The oriented special cut.

$q \equiv q' (mod\, 2p)$, just as the oriented Schubert Theorem predicts. Note that reducing all possible bottom twists implies $|p| > |q|$ for both tangles, if the two tangles that we compare each time are compatible or for only one, if they are incompatible.

We then have to compare the special cut and the palindrome cut with the standard cut. In the oriented case the special cut is the easier to see whilst the palindrome cut requires a more sophisticated analysis. Figure 29 illustrates the general case of the special cut. In order to understand Figure 29 it is necessary to also view Figure 22 for the details of this cut.

Recall that if $S = [1] + T$ then the tangle of the special cut on the knot $N([1] + T)$ is the tangle $S' = [-1] - \frac{1}{T}$. And if $F(T) = p/q$ then $F([1] + T) = \frac{p+q}{q}$ and $F([-1] - \frac{1}{T}) = \frac{p+q}{-p}$. Now, the point is that the orientations of the tangles S and S' are incompatible. Applying a $[+1]$ bottom twist to S' yields $S'' = ([-1] - \frac{1}{T}) * [1]$, and we find that $F(S'') = \frac{p+q}{q}$. Thus, the oriented rational tangles S and S'' have the same fraction and by Theorem 1 and their compatibility they are oriented isotopic and the arithmetics of Theorem 3 is straightforward.

We are left to examine the case of the palindrome cut. For this part of the proof, we refer the reader to our paper [27].

8 Strongly Invertible Links

An oriented knot or link is invertible if it is oriented isotopic to the link obtained from it by reversing the orientation of each component. We have seen (see Figure 27) that rational knots and links are invertible. A link L of two components is said to be *strongly invertible* if L is ambient isotopic to itself with the orientation of only one component reversed. In Figure 30 we illustrate the link $L = N([[2], [1], [2]])$. This is a strongly invertible link as is apparent by a 180^0 vertical rotation. This link is well-known as the Whitehead link, a link with linking number zero. Note that since $[[2], [1], [2]]$ has fraction equal to $1 + 1/(1 + 1/2) = 8/3$ this link is non-trivial via the classification of rational knots and links. Note also that $3 \cdot 3 = 1 + 1 \cdot 8$.

N([[2], [1], [2]]) = W
the Whitehead Link
F(W) = 2+1/(1+1/2) = 8/3
3 3 =·1 + 1 8 ·

Fig. 30 The Whitehead link is strongly invertible.

Fig. 31 An example of a
strongly invertible link.

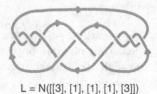

L = N([[3], [1], [1], [1], [3]])

In general we have the following. For our proof, see [27].

Theorem 8.1. *Let $L = N(T)$ be an oriented rational link with associated tangle fraction $F(T) = p/q$ of parity e/o, with p and q relatively prime and $|p| > |q|$. Then L is strongly invertible if and only if $q^2 = 1 + up$ with u an odd integer. It follows that strongly invertible links are all numerators of rational tangles of the form $[[a_1], [a_2], \ldots, [a_k], [\alpha], [a_k], \ldots, [a_2], [a_1]]$ for any integers $a_1, \ldots, a_k, \alpha$.*

See Figure 31 for another example of a strongly invertible link. In this case the link is $L = N([[3], [1], [1], [1], [3]])$ with $F(L) = 40/11$. Note that $11^2 = 1 + 3 \cdot 40$, fitting the conclusion of Theorem 8.1.

9 Applications to the Topology of DNA

DNA supercoils, replicates and recombines with the help of certain enzymes. *Site-specific recombination* is one of the ways nature alters the genetic code of an organism, either by moving a block of DNA to another position on the molecule or by integrating a block of alien DNA into a host genome. For a closed molecule of DNA a global picture of the recombination would be as shown in Figure 32, where double-stranded DNA is represented by a single line and the recombination sites are marked with points. This picture can be interpreted as $N(S + [0]) \longrightarrow N(S + [1])$, for $S = \frac{1}{[-3]}$ in this example. This operation can be repeated as in Figure 33. Note that the $[0] - [\infty]$ interchange of Figure 10 can be seen as the first step of the process.

In this depiction of recombination, we have shown a local replacement of the tangle [0] by the tangle [1] connoting a new cross-connection of the

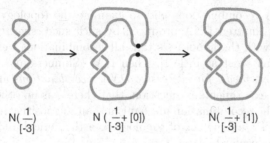

Fig. 32 Global picture of recombination.

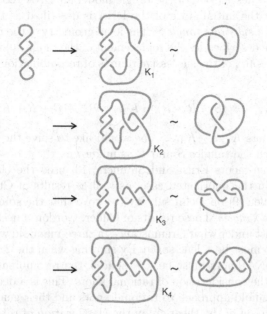

Fig. 33 Multiple recombinations.

DNA strands. In general, it is not known without corroborating evidence just what the topological geometry of the recombination replacement will be. Even in the case of a single half-twist replacement such as [1], it is certainly not obvious beforehand that the replacement will always be [+1] and not sometimes the reverse twist of [−1]. It was at the juncture raised by this question that a combination of topological methods in biology and a tangle model using knot theory developed by C.Ernst and D.W. Sumners resolved the issue in some specific cases. See [13], [48] and references therein.

On the biological side, methods of protein coating developed by N. Cozzarelli, S.J. Spengler and A. Stasiak et al. In [8] it was made possible for the first time to see knotted DNA in an electron micrograph with sufficient resolution to actually identify the topological type of these knots. The protein coating technique made it possible to design an experiment involving

successive DNA recombinations and to examine the topology of the products. In [8] the knotted DNA produced by such successive recombinations
was consistent with the hypothesis that all recombinations were of the type
of a positive half twist as in [+1]. Then D.W. Sumners and C. Ernst [13]
proposed a *tangle model for successive DNA recombinations* and showed, in
the case of the experiments in question, that there was no other topological
possibility for the recombination mechanism than the positive half twist [+1].
This constituted a unique use of topology as a theoretical underpinning for
a problem in molecular biology.

Here is a brief description of the tangle model for DNA recombination. It
is assumed that the initial state of the DNA is described as the numerator
closure $N(S)$ of a *substrate tangle* S. The local geometry of the recombination
is assumed to be described by the replacement of the tangle [0] with a specific
tangle R. The results of the successive rounds of recombination are the knots
and links

$$N(S+R) = K_1, \quad N(S+R+R) = K_2, \quad N(S+R+R+R) = K_3, \quad \ldots$$

Knowing the knots $K_1, K_2, K_3, \ldots$ one would like to solve the above system
of equations with the tangles S and R as unknowns.

For such experiments Ernst and Sumners [13] used the classification of
rational knots in the unoriented case, as well as results of Culler, Gordon,
Luecke and Shalen [9] on Dehn surgery to prove that the solutions $S + nR$
must be *rational tangles*. These results of Culler, Gordon, Luecke and Shalen
tell the topologist under what circumstances a three-manifold with cyclic fundamental group must be a lens space. By showing when the 2-fold branched
covers of the DNA knots must be lens spaces, the recombination problems
are reduced to the consideration of rational knots. This is a deep application
of the three-manifold approach to rational knots and their generalizations.

One can then apply the theorem on the classification of rational knots to
deduce (in these instances) the uniqueness of S and R. Note that, in these
experiments, the substrate tangle S was also pinpointed by the sequence of
knots and links that resulted from the recombination.

Here we shall solve tangle equations like the above under rationality
assumptions on all tangles in question. This allows us to use only the mathematical techniques developed in this paper. We shall illustrate how a sequence
of rational knots and links

$$N(S+nR) = K_n, \quad n = 0, 1, 2, 3, \ldots$$

with S and R rational tangles, such that $R = [r]$, $F(S) = \frac{p}{q}$ and $p, q, r \in \mathbb{Z}$
$(p > 0)$ *determines* $\frac{p}{q}$ *and* r *uniquely* if we know sufficiently many K_n. We
call this the "DNA Knitting Machine Analysis".

Theorem 9.1. *Let a sequence* K_n *of rational knots and links be defined by
the equations* $K_n = N(S + nR)$ *with specific integers* p, q, r $(p > 0)$, *where*

$R = [r]$, $F(S) = \frac{p}{q}$. Then $\frac{p}{q}$ and r are uniquely determined if one knows the topological type of the unoriented links $K_0, K_1, \ldots, K_N$ for any integer $N \geq |q| - \frac{p}{qr}$.

Proof. In this proof we shall write $N(\frac{p}{q} + nr)$ or $N(\frac{p+qnr}{q})$ for $N(S + nR)$. We shall also write $K = K'$ to mean that K and K' are isotopic links. Moreover we shall say for a pair of reduced fractions P/q and P/q' that q and q' are *arithmetically related relative to* P if either $q \equiv q' (mod\, P)$ or $qq' \equiv 1 (mod\, P)$. Suppose the integers p, q, r give rise to the sequence of links $K_0, K_1, \ldots$. Suppose there is some other triple of integers p', q', r' that give rise to the same sequence of links. We will show uniqueness of p, q, r under the conditions of the theorem. We shall say "the equality holds for n" to mean that $N((p + qrn)/q) = N((p' + q'r'n)/q')$. We suppose that $K_n = N((p+qrn)/q)$ as in the hypothesis of the theorem, and suppose that there are p', q', r' such that for some n (or a range of values of n to be specified below) $K_n = N((p' + q'r'n)/q')$.

If $n = 0$ then we have $N(p/q) = N(p'/q')$. Hence by the classification theorem we know that $p = p'$ and that q and q' are arithmetically related. Note that the same argument shows that if the equality holds for any two consecutive values of n, then $p = p'$. Hence we shall assume henceforth that $p = p'$. With this assumption in place, we see that if the equality holds for any $n \neq 0$ then $qr = q'r'$. Hence we shall assume this as well from now on.

If $|p + qrn|$ is sufficiently large, then the congruences for the arithmetical relation of q and q' must be *equalities over the integers*. Since $qq' = 1$ over the integers can hold only if $q = q' = 1$ or -1 we see that it must be the case that $q = q'$ if the equality is to hold for sufficiently large n. From this and the equation $qr = q'r'$ it follows that $r = r'$. It remains to determine a bound on n. In order to be sure that $|p + qrn|$ is sufficiently large, we need that $|qq'| \leq |p + qrn|$. Since $q'r' = qr$, we also know that $|q'| \leq |qr|$. Hence n is sufficiently large if $|q^2 r| \leq |p + qrn|$.

If $qr > 0$ then, since $p > 0$, we are asking that $|q^2 r| \leq p + qrn$. Hence

$$n \geq (|q^2 r| - p)/(qr) = |q| - (p/qr).$$

If $qr < 0$ then for n large we will have $|p + qrn| = -p - qrn$. Thus we want to solve $|q^2 r| \leq -p - qrn$, whence

$$n \geq (|q^2 r| + p)/(-qr) = |q| - (p/qr).$$

Since these two cases exhaust the range of possibilities, this completes the proof of the theorem.   □

Here is a special case of Theorem 7. See Figure 33. Suppose that we were given a sequence of knots and links K_n such that

$$K_n = N(\frac{1}{[-3]} + [1] + [1] + \ldots + [1]) = N(\frac{1}{[-3]} + n[1]).$$

We have $F(\frac{1}{[-3]} + n[1]) = (3n - 1)/3$ and we shall write $K_n = N([(3n - 1)/3])$. We are told that each of these rational knots is in fact the numerator closure of a rational tangle denoted

$$[p/q] + n[r]$$

for some rational number p/q and some integer r. That is, we are told that they come from a DNA knitting machine that is using rational tangle patterns. But we only know the knots and the fact that they are indeed the closures for $p/q = -1/3$ and $r = 1$. By this analysis, the uniqueness is implied by the knots and links $\{K_1, K_2, K_3, K_4\}$. This means that a DNA knitting machine $K_n = N(S + nR)$ that emits the four specific knots $K_n = N([(3n - 1)/3])$ for $n = 1, 2, 3, 4$ must be of the form $S = 1/[-3]$ and $R = [1]$. It was in this way (with a finite number of observations) that the structure of recombination in T_{n3} resolvase was determined [48].

In this version of the tangle model for DNA recombination we have made a blanket assumption that the substrate tangle S and the recombination tangle R and all the tangles $S + nR$ were rational. Actually, if we assume that S is rational and that $S + R$ is rational, then it follows that R is an integer tangle. Thus S and R neccesarily form a DNA knitting machine under these conditions. It is relatively natural to assume that S is rational on the grounds of simplicity. On the other hand it is not so obvious that the recombination tangle should be an integer. The fact that the products of the DNA recombination experiments yield rational knots and links, lends credence to the hypothesis of rational tangles and hence integral recombination tangles. But there certainly is a subtlety here, since we know that the numerator closure of the sum of two rational tangles is always a rational knot or link. In fact, it is here that some deeper topology shows that certain rational products from a generalized knitting machine of the form $K_n = N(S + nR)$ where S and R are arbitrary tangles will force the rationality of the tangles $S + nR$. We refer the reader to [13], [14], [10, 11] for the details of this approach.

References

1. M. ASAEDA, J. PRZYTYCKI AND A. SIKORA, Kauffman-Harary conjecture holds for Montesinos knots, *J. Knot Theory Ramifications* **13** (2004), no. 4, 467–477.
2. C. BANKWITZ AND H.G. SCHUMANN, Über Viergeflechte, *Abh. Math. Sem. Univ. Hamburg*, **10** (1934), 263–284.
3. S. BLEILER AND J. MORIAH, Heegaard splittings and branched coverings of B^3, *Math. Ann.*, **281**, 531–543.
4. E.J. Brody, The topological classification of the lens spaces, *Annals of Mathematics*, **71** (1960), 163–184.
5. G. BURDE, Verschlingungsinvarianten von Knoten und Verkettungen mit zwei Brücken, *Math. Zeitschrift*, **145** (1975), 235–242.
6. G. BURDE, H. ZIESCHANG, "Knots", de Gruyter Studies in Mathematics **5** (1985).

7. J.H. CONWAY, An enumeration of knots and links and some of their algebraic properties, *Proceedings of the conference on Computational problems in Abstract Algebra held at Oxford in 1967*, J. Leech ed., (First edition 1970), Pergamon Press, 329–358.

8. N. COZZARELLI, F. DEAN, T. KOLLER, M. A. KRASNOW, S.J. SPENGLER AND A. STASIAK, Determination of the absolute handedness of knots and catenanes of DNA, *Nature* , **304** (1983), 550–560.

9. M.C. CULLER, C.M. GORDON, J. LUECKE AND P.B. SHALEN, Dehn surgery on knots, *Annals of Math.*, **125** (1987), 237–300.

10. I. DARCY, Solving oriented tangle equations involving 4-plats, *J. Knot Theory Ramifications* **14** (2005), no. 8, 1007–1027.

11. I. DARCY, Solving unoriented tangle equations involving 4-plats, *J. Knot Theory Ramifications* **14** (2005), no. 8, 993–1005.

12. C.ERNST, D.W. SUMNERS, The growth of the number of prime knots, *Math. Proc. Camb. Phil. Soc.*, **102** (1987), 303–315.

13. C.ERNST, D.W. SUMNERS, A calculus for rational tangles: Applications to DNA Recombination, *Math. Proc. Camb. Phil. Soc.*, **108** (1990), 489–515.

14. C. ERNST, D.W. SUMNERS, Solving tangle equations arising in a DNA recombination model. *Math. Proc. Cambridge Philos. Soc.*, **126**, No. 1 (1999), 23–36.

15. W. FRANZ, Über die Torsion einer Überdeckung, *J. Reine Angew. Math.* **173** (1935), 245–254.

16. J.R. GOLDMAN, L.H. KAUFFMAN, Knots, Tangles and Electrical Networks, *Advances in Applied Math.*, **14** (1993), 267–306.

17. J.R. GOLDMAN, L.H. KAUFFMAN, Rational Tangles, *Advances in Applied Math.*, **18** (1997), 300–332.

18. V.F.R. JONES, A polynomial invariant for knots via von Neumann algebras, *Bull. Amer. Math. Soc. (N.S.)* **12** (1985) no. 1, 103–111.

19. V.F.R. JONES, A new knot polynomial and von Neumann algebras, *Notices Amer. Math. Soc.* **33** (1986), no. 2, 219–225.

20. L.H. KAUFFMAN, State models and the Jones polynomial, *Topology*, **26** (1987), 395–407.

21. L.H. KAUFFMAN, An invariant of regular isotopy, *Transactions of the Amer. Math. Soc.*, **318** (1990), No 2, 417–471.

22. L.H. KAUFFMAN, Knot Logic, *Knots and Applications*, Series on Knots and Everything, **2**, L.H. Kauffman ed., World Scientific, (1995).

23. L.H. KAUFFMAN, "On knots", Ann. of Math. Stud. **115**, Princeton Univ. Press, Princeton, N.J., (1987).

24. L.H. KAUFFMAN, "Formal Knot Theory", Mathematical Notes **30**, Princeton Univ. Press, Princeton, N.J., (1983), republished by Dover Publications (2006).

25. L.H. KAUFFMAN, F. HARARY, Knots and Graphs I - Arc Graphs and Colorings, *Advances in Applied Mathematics*, **22** (1999), 312–337.

26. L.H. KAUFFMAN, S. LAMBROPOULOU, On the classification of rational tangles, *Advances in Applied Math.* **33**, No. 2 (2004), 199–237.

27. L.H. KAUFFMAN, S. LAMBROPOULOU, On the classification of rational knots, *L'Enseignement Mathematiques* **49** (2003), 357–410.

28. A. KAWAUCHI, "A Survey of Knot Theory", Birkhäuser Verlag (1996).

29. A.YA. KHINCHIN, "Continued Fractions", Dover (1997) (republication of the 1964 edition of Chicago Univ. Press).

30. K. KOLDEN, Continued fractions and linear substitutions, *Archiv for Math. og Naturvidenskab*, **6** (1949), 141–196.

31. D.A. KREBES, An obstruction to embedding 4-tangles in links. *J. Knot Theory Ramifications* **8** (1999), no. 3, 321–352.

32. W.B.R. LICKORISH, "An introduction to knot theory", Springer Graduate Texts in Mathematics, **175** (1997).

33. W. MENASCO, M. THISTLETHWAITE, The classification of alternating links, *Annals of Mathematics*, **138** (1993), 113–171.

34. J.M. MONTESINOS, Revetements ramifies des noeuds, Espaces fibres de Seifert et scindements de Heegaard, *Publicaciones del Seminario Mathematico Garcia de Galdeano, Serie II, Seccion 3* (1984).
35. K. MURASUGI, " Knot Theory and Its Applications", Translated from the 1993 Japanese original by Bohdan Kurpita. Birkhuser Boston, Inc., Boston, MA, (1996). viii+341 pp.
36. C.D. OLDS, "Continued Fractions", New Mathematical Library, Math. Assoc. of Amerika, **9** (1963).
37. L. PERSON, M. DUNNE, J. DeNINNO, B. GUNTEL AND L. SMITH, Colorings of rational, alternating knots and links, (preprint 2002, unpublished).
38. V.V. PRASOLOV, A.B. SOSSINSKY, "Knots, Links, Braids and 3-Manifolds", AMS Translations of Mathematical Monographs **154** (1997).
39. K. Reidemeister, Elementare Begründung der Knotentheorie, *Abh. Math. Sem. Univ. Hamburg,* **5** (1927), 24–32.
40. K. REIDEMEISTER, "Knotentheorie" (Reprint), Chelsea, New York (1948).
41. K. REIDEMEISTER, Knoten und Verkettungen, *Math. Zeitschrift,* **29** (1929), 713–729.
42. K. REIDEMEISTER, Homotopieringe und Linsenräume, *Abh. Math. Sem. Hansischen Univ.,* **11** (1936), 102–109.
43. D. ROLFSEN, "Knots and Links", Publish or Perish Press, Berkeley (1976).
44. H. SEIFERT, Die verschlingungsinvarianten der zyklischen knotenüberlagerungen, *Abh. Math. Sem. Univ. Hamburg,* **11** (1936), 84–101.
45. J. SAWOLLEK, Tait's flyping conjecture for 4-regular graphs, *J. Combin. Theory Ser. B* **95** (2005), no. 2, 318–332.
46. H. SCHUBERT, Knoten mit zwei Brücken, *Math. Zeitschrift,* **65** (1956), 133–170.
47. L. SIEBENMANN, Lecture Notes on Rational Tangles, Orsay (1972) (unpublished).
48. D.W. SUMNERS, Untangling DNA, *Math.Intelligencer,* **12** (1990), 71–80.
49. C. SUNDBERG, M. THISTLETHWAITE, The rate of growth of the number of alternating links and tangles, *Pacific J. Math.,* **182** No. 2 (1998), 329–358.
50. P.G. TAIT, On knots, I, II, III, *Scientific Papers,* **1** (1898), Cambridge University Press, Cambridge, 273–347.
51. H.S. WALL, "Analytic Theory of Continued Fractions", D. Van Nostrand Company, Inc. (1948).

The Group and Hamiltonian Descriptions of Hydrodynamical Systems

Boris Khesin
(CIME Lecturer)

Abstract We survey applications of the Hamiltonian approach and group theory to ideal fluid dynamics and integrable systems. In particular, we review the derivations of the Landau-Lifschitz and Korteweg-de Vries equations as Euler equations on certain infinite-dimensional groups.

1 Introduction

In [1] V. Arnold showed that the Euler equation for an ideal incompressible fluid describes the geodesic flow with respect to a right-invariant metric on the corresponding group of volume-preserving diffeomorphisms. Furthermore, he suggested a general framework for Euler-type equations on arbitrary groups, which we recall below. This approach turned out to be surprisingly general: many interesting equations in mathematical physics admit descriptions as geodesic flows with respect to suitable one-sided invariant Riemannian metrics on appropriate groups. This paper is an extended version of the series of lectures given at the CIME school on Topological Fluid Mechanics. The main reference for many facts discussed below is [1, 3], see also details in [8].

B. Khesin
University of Toronto, Toronto, ON M5S 3G3, Canada
e-mail: khesin@math.toronto.edu
http://www.math.toronto.edu/khesin

R.L. Ricca (ed.), *Lectures on Topological Fluid Mechanics*,
Lecture Notes in Mathematics 1973, DOI: 10.1007/978-3-642-00837-5,
© Springer-Verlag Berlin Heidelberg 2009

2 Euler Equations and Geodesics

2.1 The Euler Equation of an Ideal Incompressible Fluid

We start with the main example, the Euler equation for an ideal fluid. Consider an incompressible fluid occupying a domain M in $\mathbb{R}^n$. The fluid motion is described by a velocity field $v(t, x)$ and a pressure field $p(t, x)$ which satisfy the classical Euler equation:

$$\partial_t v + (v, \nabla)v = -\nabla p, \tag{1}$$

where div $v = 0$ and the field v is tangent to the boundary of M. The function p is defined uniquely modulo an additive constant by the condition that v has zero divergence at any moment t. The geodesic nature of the Euler equation can be seen from the following application of the chain rule. The flow $(t, x) \to g(t, x)$ describing the motion of fluid particles is defined by its velocity field $v(t, x)$:

$$\partial_t g(t, x) = v(t, g(t, x)), \ g(0, x) = x.$$

Acceleration of particles is given by

$$\partial_t^2 g(t, x) = (\partial_t v + (v, \nabla)v)(t, g(t, x)),$$

according to the chain rule, and hence the Euler equation is equivalent to

$$\partial_t^2 g(t, x) = -(\nabla p)(t, g(t, x)).$$

The latter form of the Euler equation says that the acceleration of the flow is given by a gradient and hence it is L^2-orthogonal to the set of volume-preserving diffeomorphisms, which satisfy the incompressibility condition

$$\det(\partial_x g(t, x)) = 1.$$

More precisely, the tangent space to the set of volume-preserving diffeomorphisms (at the identity) consists of divergence-free vector fields. They are L^2-orthogonal to all gradient fields. But this is one of possible definitions of a geodesic curve: it is a curve on a manifold, whose acceleration vector is at every moment orthogonal to the manifold itself. In our case, the curve is the fluid flow, understood as a family of volume-preserving diffeomorphisms, while its acceleration, given by a gradient field, is indeed orthogonal to the set of all such diffeomorphisms. Thus the fluid motion $g(t, x)$ is a geodesic line on the set of volume-preserving diffeomorphisms of the domain M with respect to the induced L^2-metric. Note that this metric is invariant with respect to reparametrizing the fluid particles, i. e. it is *right-invariant* on the set of volume-preserving diffeomorphisms (the reparametrization of independent variable is the *right* action of a diffeomorphism). A similar equation describes an ideal incompressible fluid filling an arbitrary manifold M equipped with a volume form [1, 5]. The corresponding equation defines the geodesic flow on the group of volume-preserving diffeomorphisms of M. It turns out that the group-geodesic point of view, developed in [1] is quite fruitful for topological and qualitative understanding of the fluid motion, as well as for obtaining various quantitative results related to stability and first integrals of the Euler equation.

2.2 Geodesics on Lie Groups

Consider now the general framework proposed in [1]. Suppose that a configuration space of some physical system is a (possibly infinite-dimensional) Lie group G. For a rigid body this group is $SO(3)$, while for ideal hydrodynamics it is the group SDiff(M) of volume-preserving diffeomorphisms for an ideal fluid filling a domain M. The tangent space at the identity of the Lie group G is the corresponding Lie algebra $\mathfrak{g}$. Fix some (positive definite) quadratic form, called the energy, on the space $\mathfrak{g}$. We consider right translations of this quadratic form to the tangent space at any point of the group (the "translational symmetry" of the energy). This way the energy defines a right-invariant Riemannian metric on the group G. The geodesic flow on G with respect to this energy metric represents the extremals of the least action principle, i. e. the motions of our physical system (for a rigid body one has to consider left translations). To describe a geodesic on the *Lie group* with an initial velocity $v(0) = \xi$, we transport its velocity vector at any moment t to the identity of the group (by using the right translation). This way we obtain the evolution law for $v(t)$, given by a (non-linear) dynamical system $dv/dt = F(v)$ on the *Lie algebra* $\mathfrak{g}$ (Fig. 1).

Definition 2.1. The system on the Lie algebra $\mathfrak{g}$, describing the evolution of the velocity vector along a geodesic in a right-invariant metric on the Lie group G, is called the Euler (or Euler–Arnold) equation corresponding to this metric on G.

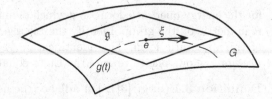

Fig. 1 The vector ξ in the Lie algebra $\mathfrak{g}$ is the velocity at the identity e of a geodesic $g(t)$ on the Lie group G.

2.3 Geodesic Description for Various Equations

A similar Arnold-type description as the geodesic flow on a Lie group can be given to a variety of conservative dynamical systems in mathematical physics. Below we list several examples of such systems to emphasize the range of applications of this approach. The choice of a group G (column 1) and an energy metric E (column 2) defines the corresponding Euler equations (column 3).

Group	Metric	Equation
$SO(3)$	$<\omega, A\omega>$	Euler top
$SO(3) \dotplus \mathbb{R}^3$	quadratic forms	Kirchhoff equations for a body in a fluid
$SO(n)$	Manakov's metrics	n-dimensional top
$\mathrm{Diff}(S^1)$	L^2	Hopf (or, inviscid Burgers) equation
Virasoro	L^2	KdV equation
Virasoro	H^1	Camassa − Holm equation
Virasoro	$\dot{H}^1$	Hunter − Saxton (or Dym) equation
$\mathrm{SDiff}(M)$	L^2	Euler ideal fluid
$\mathrm{SDiff}(M) \dotplus \mathrm{SVect}(M)$	$L^2 + L^2$	Magnetohydrodynamics
$\mathrm{Maps}(S^1, SO(3))$	H^{-1}	Landau − Lifschits equation

In some cases these systems turn out to be not only Hamiltonian, but also bi-Hamiltonian. More detailed descriptions and references can be found in the book [3].

3 Euler Equations on Groups as Hamiltonian Systems and the Binormal Equation

3.1 Hamiltonian Reformulation of the Euler Equations

The differential-geometric description of the Euler equation as a geodesic flow on a Lie group has a Hamiltonian reformulation. Fix the notation

$$E(v) = \frac{1}{2}\langle v, Av \rangle$$

for the energy quadratic form on $\mathfrak{g}$ which defines the right-invariant Riemannian metric on the group. Identify the Lie algebra and its dual with the help of this quadratic form. This identification $A : \mathfrak{g} \to \mathfrak{g}^*$ (called the *inertia operator*) allows one to rewrite the Euler equation on the dual space $\mathfrak{g}^*$.

Definition 3.1 (e. g. [3]). Let ad_v^* be (the negative of) the *coadjoint operator*, dual to the operator $[v, .]$ defining the structure of the Lie algebra $\mathfrak{g}$: for any given vector v in the Lie algebra $\mathfrak{g}$, the operator $\mathrm{ad}_v^* : \mathfrak{g}^* \to \mathfrak{g}^*$ acts on the dual space $\mathfrak{g}^*$ according to the formula

$$\langle \mathrm{ad}_v^* \xi, w \rangle := -\langle \xi, [v, w] \rangle$$

for all vectors $w \in \mathfrak{g}$ and all covectors $\xi \in \mathfrak{g}^*$. The *Euler equation on* $\mathfrak{g}^*$, corresponding to the right-invariant metric $E(v) = \frac{1}{2}\langle Av, v \rangle$ on the group, is given by the following explicit formula:

$$\frac{dm}{dt} = \mathrm{ad}_{A^{-1}m}^* m, \tag{2}$$

as an evolution of a point $m \in \mathfrak{g}^*$.

Below we explain the meaning of this operator ad^* in the cases of three different equations of mathematical physics.

Remark 3.2. It turns out that the Euler equation on $\mathfrak{g}^*$ is Hamiltonian with respect to the natural Lie-Poisson structure on the dual space [1]. Moreover, the corresponding Hamiltonian function is the energy quadratic form lifted from the Lie algebra to its dual space by the same identification: $E(m) = \frac{1}{2}\langle A^{-1}m, m \rangle$, where $m = Av$. Above we have taken this Hamiltonian equation as the *definition* of the Euler equation on the dual space $\mathfrak{g}^*$, corresponding to the geodesic flow in the right-invariant metric on the group. For left-invariant metrics the right-hand-side of the equation (2) acquires the minus sign: $-\mathrm{ad}_{A^{-1}m}^* m$.

3.2 Hamiltonian Structure of the Landau-Lifschitz Equation

Consider a vector-valued function $L = L(\theta)$ on the circle with values in $\mathbb{R}^3$.

Definition 3.3. The *Landau-Lifschitz equation*, called also the *Heisenberg magnetic chain equation*, is defined as the following evolution equation on the function L:

$$\partial_t L = L \times \partial_\theta^2 L \tag{3}$$

This equation can be viewed as the Euler equation for a certain *left-invariant* metric on the loop group $LSO(3) = C^\infty(S^1, SO(3))$ of matrix-valued functions on the circle with values in the group $SO(3)$.

Definition 3.4. Consider the Lie algebra $\mathfrak{g} = L\mathfrak{so}(3)$ of smooth functions on the circle with values in 3×3 skew-symmetric matrices and the pointwise commutator. The elements of this algebra are naturally identified with vector-valued functions with values in $\mathbb{R}^3$ and whose commutator is given by the cross-product of their values:

$$[X(\theta), Y(\theta)] = X(\theta) \times Y(\theta).$$

Furthermore, regard the dual space $\mathfrak{g}^*$ as the same space of vector-valued functions with the natural pairing between the spaces:

$$\langle X(\theta), L(\theta) \rangle = \int_{S^1} (X(\theta), L(\theta)) \, d\theta$$

for $X \in \mathfrak{g}$, $L \in \mathfrak{g}^*$ and where we used the Euclidean product $(\,,\,)$ in $\mathbb{R}^3$.

Theorem 3.5. *The Landau-Lifschitz equation is the Euler equation corresponding to the left-invariant H^{-1}-metric on the loop group $LSO(3) = C^\infty(S^1, SO(3))$:*

$$E(X) = \frac{1}{2} \int_{S^1} (\partial_\theta^{-1} X, \partial_\theta^{-1} X) \, d\theta \,.$$

Here ∂_θ^{-1} is the integration operator inverse to the differentiation $\partial_\theta := \partial/\partial\theta$ on the space of functions with zero mean on the circle.

Proof. By using integration by parts we rewrite the energy in the form

$$E(X) = \frac{1}{2} \int_{S^1} (\partial_\theta^{-1} X, \partial_\theta^{-1} X) \, d\theta = -\frac{1}{2} \int_{S^1} (\partial_\theta^{-2} X, X) \, d\theta \,,$$

which corresponds to the inertia operator $A = \partial_\theta^{-2}$ (also defined on functions with zero mean). For the Euler equation we need the inverse of the inertia operator, that is even easier to define:

$$A^{-1}(X) = \partial_\theta^2 X \,.$$

Finally, note that the operator ad^* has the form of the cross product, just like the commutator in the Lie algebra:

$$\mathrm{ad}_v^* L = v \times L.$$

Indeed,

$$\langle \mathrm{ad}_v^* L, w \rangle = -\langle L, [v, w] \rangle = -\int_{S^1} (L, v \times w) \, d\theta = -\int_{S^1} (L \times v, w) \, d\theta = \langle v \times L, w \rangle.$$

Combining the above we obtain the Euler equation for the left-invariant H^{-1}-metric (i. e. with the minus sign, see Remark 3.2). It has the Landau-Lifschitz form:

$$\partial_t L = -\mathrm{ad}_{A^{-1}(L)}^* L = L \times \partial_\theta^2 L \,,$$

as required.   □

   In a similar way one can derive analogues of the Landau-Lifschitz equation for the loops in matrices of higher order. However, the 3×3 case discussed

Fig. 2 The curve evolution
in the direction of the
binormal.

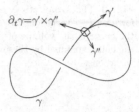

above has several equivalent reformulations and specific 3D features. The
most straightforward reformulation of the Landau-Lifschitz equation is in
the form of the binormal equation. Namely, let $\gamma : S^1 \to \mathbb{R}^3$ be a closed
immersed curve in $\mathbb{R}^3$ with an arc-length parameter θ.

Definition 3.6. The *binormal* or *filament equation* is the evolution of the
closed curve γ according to the following law:

$$\partial_t \gamma = \gamma' \times \gamma'', \tag{4}$$

see Fig. 2. (Here, γ' denotes the derivative $\frac{\partial}{\partial \theta}\gamma$.)

We can see that the binormal equation is equivalent to the Landau-
Lifschitz equation by setting $L = \gamma'$: we obtain

$$\partial_t L = \partial_t \gamma' = \gamma'' \times \gamma'' + \gamma' \times \gamma''' = \gamma' \times \gamma''' = L \times L''.$$

There are several equivalent forms for the binormal equation (4), which
we discuss in the next section. Note that if θ is not arc-length, then the
corresponding Hamiltonian equation becomes

$$\partial_t \gamma = k(\theta, t)(\gamma' \times \gamma''), \tag{5}$$

where $k(\theta, t)$ is the curvature of the curve at the point θ at time t. This
equation is also called Da Rios or LIA (Localised Induction Approximation)
equation, see a nice survey [13].

3.3 Properties of the Binormal Equation

Even without the reference to the Landau-Lifschitz equation, the binormal
equation is Hamiltonian with respect to a nice symplectic structure which
exists on the space of knots in $\mathbb{R}^3$ with a fixed volume form μ. Consider the
space $\mathcal{C}$ whose points are immersed closed curves in $\mathbb{R}^3$. A tangent vector a to
the space $\mathcal{C}$ at γ is an infinitesimal variation of the curve γ, that is a normal
vector field $a(\theta)$ attached to γ.

Definition 3.7 ([9]). The *Marsden-Weinstein symplectic structure* is the following 2-form ω_{MW} on the space $\mathcal{C}$. The value of the symplectic form ω_{MW}

Fig. 3 The symplectic
form on two variations of
the curve is the volume of
the collar spanned by these
variations.

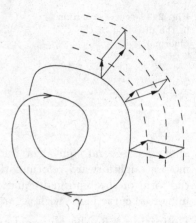

on two tangent vectors a and b to the space $\mathcal{C}$ at γ is given by the oriented
volume of the collar spanned along γ by the vectors $a(\theta)$ and $b(\theta)$:

$$\omega_{MW}(a,b) := \int_{S^1} \mu(a(\theta), b(\theta), \gamma'(\theta)) d\theta = \int_\gamma \iota_b \iota_a \, \mu,$$

see Fig. 3.

Define the following Hamiltonian function H on the space of immersed
closed curves by assigning to a path γ its length:

$$H(\gamma) = \int_{S^1} \sqrt{(\gamma'(\theta), \gamma'(\theta))} \, d\theta.$$

Theorem 3.8 (e. g. [4]). *The Hamiltonian equation corresponding to the
length Hamiltonian function H with respect to the Marsden-Weinstein symplectic structure is given by the binormal equation.*

Notice that the binormal vector $\gamma' \times \gamma''$ is orthogonal to γ', and hence the
evolution of the curve γ does not change the length of γ, i. e. the Hamiltonian
itself is indeed a first integral of the binormal equation.

Proof. In order to see why the length functional indeed gives the binormal
equation we notice that for the functional $H(\gamma)$ in the arc-length parametriza-
tion the variational derivative (i. e. the "gradient" of H in the space of curves)
is $\delta H/\delta \gamma = -(\text{length of } \gamma)^{-1} \cdot \gamma''$. This can be thought of as a normal to γ
vector field. In turn, the Marsden-Weinstein symplectic structure at γ can
be regarded as the symplectic structure averaged over all two-dimensional
planes normal to γ. Then to obtain the corresponding Hamiltonian field v_H,
which is the skew-gradient of H, we have to rotate the above gradient field
$\delta H/\delta \gamma$ by $\pi/2$ in all such normal planes. Given θ, the corresponding rotation

operator in the plane normal to the tangent vector $\gamma'(\theta)$ is $J_\gamma(*) = \gamma' \times *$. Thus we obtain:

$$v_H = J_\gamma(\delta H/\delta\gamma) = -(\text{length of } \gamma)^{-1} \cdot (\gamma' \times \gamma'').$$

$\square$

The Marsden-Weinstein symplectic structure is closely related to the Hamiltonian description of the ideal incompressible fluid filling $\mathbb{R}^3$, see the discussion in Section 5.2.

Remark 3.9. The binormal equation $\partial_t\gamma = \gamma' \times \gamma''$ can be rewritten in the Frenet frame for the curve γ as evolution equations for the curvature $k(\theta, t)$ and torsion $\tau(\theta, t)$, see [14]. By passing to the new functions "velocity" $v := \tau$ and "energy density" $\rho := k^2$ one obtains the equations of 1-dimensional barotropic fluid

$$\begin{cases} \partial_t v = -(v, \nabla)v - \nabla h(\rho)/\rho \\ \partial_t\rho + \text{div}(\rho \cdot v) = 0, \end{cases}$$

for a specific function $p = h(\rho)$, relating density and pressure.

Another important feature of the binormal equation was observed by Hasimoto [6]. He found the transformation

$$\psi(\theta, t) = k(\theta, t)\exp\left(i \cdot \int_0^\theta \tau(\eta, t)d\eta\right),$$

which sends the binormal equation to the non-linear Schrödinger equation:

$$-i\,\partial_t\psi = \psi'' + \frac{1}{2}|\psi|^2\psi$$

for a complex-valued wave function $\psi : S^1 \to \mathcal{C}$. The latter equation is known to be a completely integrable (bi-Hamiltonian) infinite-dimensional system, and hence so is the binormal or Landau-Lifschitz equations.

4 The KdV-Type Equations as Euler Equations

4.1 The Virasoro Algebra and the KdV Equation

Definition 4.1. The Virasoro algebra $vir = \text{Vect}(S^1) \oplus \mathbb{R}$ is the vector space of pairs which consist of a smooth vector field on the circle and a number. This space is equipped with the following commutation operation:

$$[(f(x)\frac{\partial}{\partial x}, a), (g(x)\frac{\partial}{\partial x}, b)] = \left((f'(x)g(x) - f(x)g'(x))\frac{\partial}{\partial x}, \int_{S^1} f'(x)g''(x)\,dx\right),$$

for any two elements $(f(x)\partial/\partial x,\ a)$ and $(g(x)\partial/\partial x, b)$ in vir. The bilinear skew-symmetric expression $c(f, g) := \int_{S^1} f'(x)g''(x)dx$ is called the *Gelfand–Fuchs 2-cocycle*.

There exists a Virasoro group, an extension of the group of smooth diffeomorphisms of the circle, whose Lie algebra is the Virasoro algebra vir. Fix the L^2-energy quadratic form in the Virasoro Lie algebra:

$$E(f(x)\frac{\partial}{\partial x},\ a) = \frac{1}{2}\left(\int_{S^1} f^2(x)\ dx + a^2\right).$$

As before, we equip the Virasoro group with a (right-invariant) Riemannian metric and consider the corresponding Euler equation, i. e. the equation of the geodesic flow generated by this metric on the Virasoro group.

Theorem 4.2. [12] *The Euler equation corresponding to the geodesic flow (for the above right-invariant L^2-metric) on the Virasoro group is a one-parameter family of the Korteweg–de Vries (KdV) equations:*

$$\partial_t u + u\partial_x u + c\partial_x^3 u = 0; \quad \partial_t c = 0$$

on a time-dependent function u on S^1. Here c is a (constant) parameter, the "depth" of the fluid.

Proof. The dual space vir^* can be identified with the set of pairs

$$\{(u(x)(dx)^2, c)|\ u(x)\ \text{is a smooth function on}\ S^1,\ c \in \mathbb{R}\}.$$

Indeed, it is natural to contract the quadratic differentials $u(x)(dx)^2$ with vector fields on the circle, while the constants are to be paired between themselves:

$$\langle (u(x)(dx)^2, c), (v(x)\frac{\partial}{\partial x},\ a)\rangle = \int_{S^1} u(x)\cdot v(x)\ dx + a\cdot c.$$

The coadjoint action of a Lie algebra element $(f\,\partial/\partial x, a) \in vir$ on an element $(u(x)(dx)^2, c)$ of the dual space vir^* is

$$\mathrm{ad}^*_{(f\,\partial/\partial x,a)}(u(dx)^2, c) = -(2(\partial_x f)u + f\partial_x u + c\partial_x^3 f,\ 0).$$

It is obtained from the identity

$$\langle \mathrm{ad}^*_{(f\frac{\partial}{\partial x},\ a)}(u(dx)^2, c), (g\frac{\partial}{\partial x},\ b)\rangle = -\langle (u(dx)^2, c), [(f\frac{\partial}{\partial x},\ a), (g\frac{\partial}{\partial x},\ b)]\rangle,$$

which holds for every pair $(g\frac{\partial}{\partial x},\ b) \in vir$. The quadratic energy functional E on the Virasoro algebra determines the "tautological" inertia operator A :

$vir \to vir^*$, which sends a pair $(u(x)\partial/\partial x, c) \in vir$ to $(u(x)(dx)^2, c) \in vir^*$. The corresponding Euler equation for the right-invariant metric on the group (according to the general formula (2)) is given by

$$\frac{\partial}{\partial t}(u(dx)^2, c) = \mathrm{ad}^*_{A^{-1}(u(dx)^2, c)}(u(dx)^2, c).$$

Making use of the explicit formula for the Virasoro coadjoint action ad^* for

$$(f\,\partial/\partial x, a) = A^{-1}(u(dx)^2, c) = (u\,\partial/\partial x, c),$$

we obtain the required Euler equation:

$$\partial_t u = -2(\partial_x u)u - u\partial_x u - c\partial_x^3 u = -3u\partial_x u - c\partial_x^3 u, \quad \partial_t c = 0.$$

The coefficient c is preserved in time, and the function u satisfies the KdV equation. $\square$

4.2 Similar Equations and Conservation Laws

If we change the metric on the Virasoro group, other interesting equations arise from the same scheme. The Euler equation on the Virasoro group with respect to the right-invariant H^1-metric gives the Camassa-Holm equation:

$$\partial_t u - \partial_{xxt} u = -3u\partial_x u + 2(\partial_x u)\partial_{xx} u + u\partial_{xxx} u + c\partial_x^3 u,$$

see [10]. Similarly, the homogeneous $\dot{H}^1$-metric gives the Hunter-Saxton equation (an equation in the Dym hierarchy):

$$\partial_{xxt} u = -2(\partial_x u)\partial_{xx} u - u\partial_{xxx} u,$$

see [7].

Remark 4.3. It turns out that all these three equations (KdV, CH, and HS) are bi-Hamiltonian systems, and hence admit an infinite family of conservation laws. The corresponding Hamiltonian (or Poisson) structures are naturally related to the Virasoro algebra. For instance, for the KdV equation these conserved quantities can be expressed in the following way. Consider the KdV equation on $(u(x)(dx)^2,\ c)$ as an evolution of Hill's operator $c\frac{d^2}{dx^2} + u(x)$. The monodromy $M(u)$ of this operator is a 2×2-matrix with $\det M(u) = 1$. Look at the following function of the monodromy for a family of Hill's operators:

$$h_\lambda\left(\frac{d^2}{dx^2} + u(x) - \lambda^2\right) := \log(\mathrm{trace}\, M(u - \lambda^2)),$$

where $M(u - \lambda^2)$ is the monodromy of the operator $\frac{d^2}{dx^2} + u(x) - \lambda^2$. Now, the expansion of the function h_λ in λ as $\lambda \to \infty$ produces the first integrals of the KdV equation:

$$h_\lambda \approx 2\pi\lambda - \sum_{n=1}^{\infty} c_n h_{2n-1} \lambda^{1-2n},$$

where

$$h_1 = \int_{S^1} u(x)\,dx, \quad h_3 = \int_{S^1} u^2(x)\,dx, \quad h_5 = \int_{S^1} \left(u^3(x) - \frac{1}{2}(u_x(x))^2 \right)\,dx \ldots$$

and c_n are constants. Note that the trace of a monodromy is a so called Casimir function for the Virasoro algebra, i. e an invariant with respect to the action of variable changes on Hill's operators. The general theory states that the coefficients in the Casimir expansion provide a hierarchy of conserved charges for a bi-Hamiltonian systems, see more details on this and other equations in [7].

5 Hamiltonian Structure of the Euler Equations for an Incompressible Fluid

5.1 The Euler Hydrodynamics as a Hamiltonian Equation

The geodesic properties of the Euler equation can also be described within the Hamiltonian framework. The latter turns out to be useful to establish numerous conservation laws for the ideal fluid. The Lie algebra $\mathfrak{g} = \mathrm{SVect}(M)$ for an ideal incompressible fluid filling a compact manifold M with a volume form μ consists of divergence-free vector fields on M. The main example for us will be a domain M in the Euclidean space $\mathbb{R}^n$. Elements of the corresponding dual space $\mathfrak{g}^*$ can be thought of as 1-forms u on M defined modulo function differentials. In other words, such 1-forms can be regarded as linear functionals on vector fields by means of the pairing

$$< u, \xi > = \int_M u(\xi)\,\mu$$

for any divergence-free field ξ. One can see that if u is an exact 1-form, i. e. if u is the differential df of a smooth function f, the corresponding pairing with a divergence-free field is necessarily zero. This is why the dual space $\mathfrak{g}^*$ is actually the quotient $\mathfrak{g}^* = \Omega^1(M)/d\Omega^0(M)$ of all 1-forms on M modulo exact 1-forms. Elements of this quotient are called cosets and denoted by

$[u] := \{u + df \mid u \in \Omega^1(M),\ f \in C^\infty(M)\}$. The energy quadratic form on vector fields is their L^2-energy

$$E(\xi) = \int_M (\xi, \xi)\, \mu,$$

with respect to the fixed Riemannian metric $(\ ,\ )$ on the manifold M. The corresponding inertia operator

$$A : \mathfrak{g} = \mathrm{SVect}(M) \to \mathfrak{g}^* = \Omega^1(M)/d\Omega^0(M)$$

sends a vector field ξ to the coset $[u]$ of the 1-form u which is obtained from the field ξ by "raising indices" with help of the metric: $u(\,.\,) = (\xi, \,.\,)$. Since the Riemannian metric is non-degenerate, so is the map $A : \xi \mapsto [u]$. In $\mathbb{R}^n$ with the standard Euclidean metric, the inertia operator A maps a vector field $\xi = \sum \xi_i \partial/\partial x_i$ to the 1-form $u_\xi = \sum \xi_i dx_i$. It turns out that the operator ad^*_ξ of the coadjoint action has a natural geometric description.

Proposition 5.1. *The coadjoint action of $S\mathrm{Vect}(M)$ on $\mathfrak{g}^* = \Omega^1(M)/d\Omega^0(M)$ is given by the negative of the Lie derivative of the 1-form u (or of its coset $[u]$):*

$$\mathrm{ad}^*_\xi \ : \ [u] \mapsto -[L_\xi u]$$

for any divergence free vector field ξ on M.

Proof. The commutator in the Lie algebra $\mathfrak{g} = \mathrm{SVect}(M)$is (minus) the commutator of two vector fields on M:

$$[\xi, \eta]_\mathfrak{g} = -[\xi, \eta]_M = -L_\xi \eta.$$

Then by definition of the coadjoint operator

$$\langle \mathrm{ad}^*_\xi u, \eta \rangle = -\langle u, [\xi, \eta]_\mathfrak{g} \rangle = \langle u, L_\xi \eta \rangle = \int_M u(L_\xi \eta)\, \mu = -\int_M (L_\xi u)(\eta)\, \mu = -\langle L_\xi u, \eta \rangle,$$

where in the transfer of the Lie derivative L_ξ from the vector field η to the 1-form u we used that the field ξ is divergence-free, and hence the term containing the Lie derivative $L_\xi \mu$ vanishes. The above implies that $\mathrm{ad}^*_\xi = -L_\xi$. Note also that the Lie derivative operator ad^* are well-defined on the cosets, since changes of coordinates commute with taking d. $\square$

Now we can combine the above expressions of the coadjoint and inertia operators to obtain the Hamiltonian form of the Euler hydrodynamics equation. The Euler equation (2) for the right-invariant metric has the form

$$\frac{d}{dt}[u] = \mathrm{ad}^*_\xi [u] = -L_\xi[u], \tag{6}$$

where $[u]$ is the coset of the 1-form u related to the divergence-free field ξ with the help of the metric: $u(\,.\,) = (\xi, \,.\,)$. Passing from the equation on the cosets $[u]$ to the equation on 1-forms u themselves we get the same equality as above, which holds only modulo adding the differential of a function:

$$\partial_t u = -L_\xi u - d\tilde{p} \tag{7}$$

for some time-dependent function $\tilde{p}$. Now we can return to vector fields by lowering indices (i. e. by applying the inverse inertia operator A^{-1} to both sides of the latter equation), and we obtain

$$\partial_t \xi = -(\xi, \nabla)\xi - \nabla p,$$

the original Euler equation (1) for an ideal incompressible fluid on M, where p is interpreted as the pressure function. (Here we use the fact that "raising indices" sends the covariant derivative $(\xi, \nabla)\xi$ to the Lie derivative $L_\xi u$ up to a full differential, see e. g. [3]). This description of the dual space and the Euler hydrodynamics equation allows one to describe certain first integrals of the fluid motion practically without calculations.

Remark 5.2. In the case of a 3-dimensional manifold M, consider the functional

$$I(u) = \int_M u \wedge du.$$

One can see that for any smooth function f,

$$I(u + df) = \int_M (u + df) \wedge d(u + df) = \int_M u \wedge du = I(u) =: I([u]),$$

i.e. this functional is well defined on the space of cosets $\Omega^1(M)/d\Omega^0(M)$. Since the integral I is defined in a coordinate-free way on this space, it is invariant under the action of the group SDiff(M) on 1-forms by coordinate changes. The latter implies that the integral I is a Casimir function, i. e. in particular, it is a first integral of the Euler equation for whatever Riemannian metric on M we choose. Furthermore, if the 1-form u and a divergence-free field ξ are related by means of the Riemannian metric $u(.) = (\xi, .)$, a short direct calculation allows one to rewrite $I(u)$ in the form

$$I(\xi) = \int_M (\xi, \operatorname{curl} \xi)\, \mu.$$

The latter integral has a natural geometric meaning of *helicity* of the vector field $\operatorname{curl} \xi$: The helicity of a vector field has a topological interpretation via "average linking number" of the trajectories of this field (see [11, 2] for details). Similarly, in the case of a two-dimensional domain $M \subset \mathbb{R}^2$, one can interpret geometrically the invariance of the enstrophy integrals

$$I_k(\xi) = \int_M (\operatorname{curl} \xi)^k d^2 x,$$

where $\operatorname{curl} \xi := \partial \xi_1/\partial x_2 - \partial \xi_2/\partial x_1$ is the *vorticity function* on $\mathbb{R}^2$.

5.2 The Space of Knots and the Dual of the Lie Algebra of Divergence-Free Vector Fields

The geometric characterization of the dual to the Lie algebra of divergence-free vector fields allows one to define a symplectic structure on the space of knots in 3D. Consider a (possibly knotted) embedded closed curve γ in Euclidean space $\mathbb{R}^3$. Associate to this curve the following linear functional on the space $\mathrm{SVect}(\mathbb{R}^3)$ of divergence-free fields in $\mathbb{R}^3$, that is an element of the dual space $\mathrm{SVect}(\mathbb{R}^3)^*$. Take a compact oriented surface $S \subset \mathbb{R}^3$ such that the oriented boundary of S coincides with the knot γ: $\partial S = \gamma$. Such a surface is called a *Seifert surface* for γ, see Fig. 4. Then to any vector field ξ in $\mathbb{R}^3$ we can associate its flux through the surface S. If the vector field ξ is divergence-free, the flux does not depend on the choice of the surface S, so that the curve γ defines a functional on the Lie algebra $\mathrm{SVect}(\mathbb{R}^3)$:

$$\langle \gamma, \xi \rangle = Flux(\xi)|_S = \int_S \iota_\xi \mu, \tag{8}$$

where $\mu = d^3x$ is the standard volume form in $\mathbb{R}^3$.

Remark 5.3. Although the functional on the Lie algebra $\mathrm{SVect}(\mathbb{R}^3)$ defined by the curve γ does not lie in the *smooth part* of the dual of $\mathrm{SVect}(\mathbb{R}^3)$, we can associate to it (the coset of) *singular*, rather than smooth, 1-forms on $\mathbb{R}^3$. Namely, given a Seifert surface S we consider the "δ-type" 1-form u_S supported on S, whose integral over any closed curve σ in $\mathbb{R}^3$ counts (with appropriate multiplicities) the intersections of σ with the surface S. Note that although the 1-form u_S depends on the choice of the surface S, its coset $[u_S]$ does not: the choice of another Seifert surface $\widetilde{S}$ changes the 1-form u_S by a full differential. This coset $[u_S]$ belongs to a *completion* of the dual space $\mathfrak{g}^* = \Omega^1(\mathbb{R}^3)/d\Omega^0(\mathbb{R}^3)$. Note that the latter space of all 1-forms modulo exact 1-forms is isomorphic to the space $Z^2(\mathbb{R}^3)$ of all closed 2-forms in $\mathbb{R}^3$: the exterior derivative d takes any such coset of 1-forms to a closed 2-form without any loss of information, since $H^1(\mathbb{R}^3) = 0$. Geometrically, we associate to the curve γ a singular closed 2-form ω_γ in $\mathbb{R}^3$ which is the δ-type 2-form (called the de Rham current) supported on γ. One can easily see that $d[u_S] = \omega_\gamma$ as currents, which exactly manifests the relation between a knot and its Seifert surface. This way a knot γ in $\mathbb{R}^3$ can be seen as an element

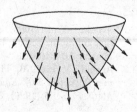

Fig. 4 A Seifert surface for the unknot defines the flux functional on vector fields.

of the dual space $\mathfrak{g}^*$ to the Lie algebra $\mathfrak{g} = \text{SVect}(\mathbb{R}^3)$ of divergence-free vector fields on $\mathbb{R}^3$. From this viewpoint, the coadjoint orbit through γ is the equivalence class of the knot γ under various (volume-preserving) isotopies. This leads to the curious observation that all *knot invariants* constitute a part of Casimirs, i.e. the coadjoint invariants of the whole Lie group $\text{SDiff}(\mathbb{R}^3)$, see e. g. [2, 3].

One can show that the Marsden-Weinstein symplectic structure on the space of embedded curves containing γ discussed in Section 3.3 coincides with the natural Kirillov–Kostant symplectic structure on the coadjoint orbit of the group $\text{SDiff}(\mathbb{R}^3)$ passing through the curve γ, where the curve is regarded as an element of the dual space to $S\text{Vect}(\mathbb{R}^3)$. The Euler dynamics on such curves γ is called the *dynamics of vorticity lines* in $\mathbb{R}^3$. Although this dynamics is governed by the non-local interaction law in terms of vorticity functions, the Helmholtz equation, one can truncate this equation by keeping only local terms. This leads to the Localized Induction Approximation (LIA), or the binormal equation. This truncation turns out to be integrable. For $n = 2$ the corresponding local approximation recover the *dynamics of point vortices*, much studied in the literature. The same construction works for any three-dimensional manifold equipped with a volume form. In higher dimensions, a similar construction allows one to define a symplectic structure on the space of closed immersed codimension 2 submanifolds of M. One can also prove its equivalence with the Kirillov–Kostant symplectic structure on ("singular") coadjoint orbits for $\text{SDiff}(M)$, which are linear functionals on divergence-free vector fields in M, represented by "fluxes through" these submanifolds.

Acknowledgments

This paper presents the lecture notes for author's CIME minicourse given in the Summer 2001. I am thankful to the organizers of this CIME school, and in particular to Professor Renzo Ricca. I am also grateful to the MPI in Bonn, where these notes were completed, and to Robert Wendt for the figures to the notes. The present work was partially sponsored by an NSERC research grant.

References

1. Arnold, V. I. (1966) Sur la géométrie différentielle des groupes de Lie de dimension infinie et ses applications à l'hydrodynamique des fluides parfaits. *Ann. Inst. Fourier* **16**, 316–361.
2. Arnold, V. I. (1973) The asymptotic Hopf invariant and its applications. *Proc. Summer School in Diff. Equations at Dilizhan,* Erevan (in Russian); English transl. : *Sel. Math. Sov.* **5** (1986), 327–345.

3. Arnold, V. I. & Khesin, B. A. (1998) Topological methods in hydrodynamics. *Applied Mathematical Sciences, vol. 125, Springer-Verlag, New York*, pp. xv+374.

4. Calini, A. (2000) Recent developments in integrable curve dynamics. In *Geom. Approaches to Diff. Equations*; Australian Math. Soc. Lect. Notes Ser., **15**, Cambridge University Press, 56–99.

5. Ebin, D. & Marsden, J. (1970) Groups of diffeomorphisms and the notion of an incompressible fluid. *Ann. of Math. (2)* **92**, 102–163.

6. Hasimoto, H. (1972) A soliton on a vortex filament, *J. Fluid Mechanics* **51**, 477–485.

7. Khesin, B. & Misiołek, G. (2003) Euler equations on homogeneous spaces and Virasoro orbits. *Advances in Math.* **176**, 116–144.

8. Khesin, B. & Wendt, R. (2007) The geometry of infinite-dimensional groups. *Ergebnisse der Mathematik und ihrer Grenzgebiete. 3. Folge*, Springer-Verlag, to appear.

9. Marsden, J. E., Weinstein, A. (1983) Coadjoint orbits, vortices, and Clebsch variables for incompressible fluids, *Physica D* **7**, 305–323.

10. Misiołek, G. (1998) A shallow water equation as a geodesic flow on the Bott-Virasoro group. *J. Geom. Phys.* **24:3**, 203–208; Classical solutions of the periodic Camassa-Holm equation. *Geom. Funct. Anal.* **12:5** (2002), 1080–1104.

11. Moffatt, H. K. (1969) The degree of knottedness of tangled vortex lines, *J. Fluid. Mech.* **106**, 117–129

12. Ovsienko, V. Yu. & Khesin, B. A. (1987) Korteweg-de Vries super-equation as an Euler equation. *Funct. Anal. Appl.* **21:4**, 329–331.

13. Ricca, R. L. (1996) The contributions of Da Rios and Levi-Civita to asymptotic potential theory and vortex filament dynamics, *Fluid Dynam. Res.* **18:5**, 245–268.

14. Turski, L. A. (1981) Hydrodynamical description of the continuous Heisenberg chain, *Canad. J. Phys.* **59:4**, 511–514.

Singularities in Fluid Dynamics and their Resolution

H.K. Moffatt
(CIME Lecturer)

Abstract Three types of singularity that can arise in fluid dynamical problems will be distinguished and discussed. These are: (i) singularities driven by boundary motion in conjunction with viscosity (e.g. corner singularities, or the Euler-disc finite-time singularity); (ii) free-surface (cusp) singularities associated with surface-tension and viscosity; (iii) interior point singularities of vorticity associated with intense vortex stretching. The singularities of types (i) and (ii) are now well known, and mechanisms by which the singularities may be resolved are clear. The question of existence of singularities of type (iii) is still open; current evidence for and against will be discussed.

1 Introduction

This paper is concerned with some examples of singularities, i.e unbounded local behaviour of the velocity field or its derivatives, in the flow of an incompressible fluid, and the manner in which such singularities must in practice be resolved. Singularities may be associated with the geometry of the fluid boundary or with some singular feature of the motion of the boundaries; they may arise spontaneously at a free surface as a result of viscous stresses and despite the smoothing action of surface tension; or they may conceivably occur at interior points of a fluid due to unbounded vortex stretching at high (or infinite) Reynolds number. In the last case, we are up against the unsolved and extremely challenging 'finite-time-singularity' problem for the Euler and/or Navier-Stokes equations. The question of existence of finite-time singularities is still open. Solution of this problem would have far-reaching

H.K. Moffatt
Department of Applied Mathematics and Theoretical Physics,
University of Cambridge Wilberforce Road, Cambridge CB3 0WA, UK
e-mail: H.K.Moffatt@damtp.cam.ac.uk

R.L. Ricca (ed.), *Lectures on Topological Fluid Mechanics*,
Lecture Notes in Mathematics 1973, DOI: 10.1007/978-3-642-00837-5,
© Springer-Verlag Berlin Heidelberg 2009

consequences for our understanding of the smallest-scale features of turbulent flow.

I have discussed some aspects of each of these problems previously (Moffatt 2001). In the present paper however, I shall place greater emphasis on the means by which each type of singularity may be resolved, and I shall indicate the nature of some recent advances.

2 Boundary-Driven Singularities

The simplest, and most prototypical, example of a boundary-driven singularity is given by the 'paint-scraper' problem of G.I. Taylor (1960) illustrated in figure 1. Fluid is contained in the corner between a fixed plate $\theta = \alpha$ and a plate $\theta = 0$ which moves parallel to itself with velocity U. Here, the singularity is imposed firstly through the geometrical singularity (the curvature of the fluid boundary being infinite at O), and secondly through the imposed discontinuity of boundary velocity at O.

Inertial forces are negligible in a region $r \ll \nu/|U|$ near the corner. Taylor's well-known similarity solution for the streamfunction $\psi(r, \theta)$ in this region takes the form $\psi(r, \theta) = -Urf(\theta)$, where

$$f(\theta) = \frac{(\alpha^2 - k\theta)\sin\theta - \theta\sin^2\alpha\cos\theta}{\alpha^2 - \sin^2\alpha}, \quad k = \tfrac{1}{2}(2\alpha - \sin 2\alpha). \quad (1)$$

This yields a velocity field which is finite throughout the fluid domain. However, the stress field has a non-integrable $O(r^{-1})$ singularity as $r \to 0$. In particular, the pressure field has the form

$$p = p_0 - \frac{2\mu U}{r}\frac{\alpha\sin\theta + \sin\alpha\sin(\alpha - \theta)}{\alpha^2 - \sin^2\alpha}, \quad (2)$$

where μ is the dynamic viscosity of the fluid. This tends to $\pm\infty$ as $r \to 0$ according as $U <$ or > 0. In either case, the force F (per unit length) required to hold the scraper in position is infinite. This just indicates that there is something wrong with the proposed solution!

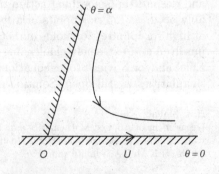

Fig. 1 Flow in a corner driven by tangential motion of one boundary. The pressure and stress are singular at the corner.

The resolution of the singularity is simple to describe in physical terms, but, so far as I know, no mathematical description is as yet available. Consider first the case $U < 0$, when $p \to +\infty$ at O. If F is large but finite, then, as recognised by Taylor, we must allow for a small gap between the two plates (figure 2), which is of the same order of magnitude as the thickness of the layer of paint spread on the plate in the paint-scraper context. If the length of the scraper (in the r-direction) is L, then the force F is of order $2\mu|U|\ln(L/\delta)$, and so δ has order of magnitude

$$\delta \sim L\exp(-cF/\mu|U|)\,, \tag{3}$$

where c is a constant of order unity.

If $U > 0$, so that according to (2), $p \to -\infty$ at O, then cavitation must presumably occur in a δ-neighbourhood of O (figure 3). Within the cavity, the pressure p equals the vapour pressure p_v, and from (2), δ is given in order of magnitude by

$$\delta \sim (p_0 - p_v)/\mu U\,, \tag{4}$$

p_0 being interpreted as the pressure far from the corner. The force F is then rendered finite:

$$F \sim 2\mu U\ln(L\mu U/(p_0 - p_v))\,. \tag{5}$$

In both cases, verification of this description requires solution of a difficult mixed boundary-value problem involving determination of the shape of the free surface on which the tangential stress is zero and the normal stress is constant.

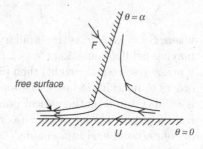

Fig. 2 Stress resolution when $U < 0$: for finite force applied to the stationary plate, a small gap is necessarily present, and a small amount of fluid leaks through the gap.

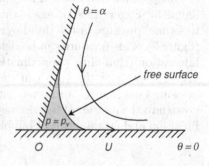

Fig. 3 Stress resolution when $U > 0$: the liquid in the corner must cavitate, the shape of the free surface being determined by the condition that the vapour pressure is constant in the cavity.

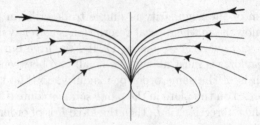

Fig. 4 Cusp formation on the free surface Γ of a viscous fluid, the flow being induced by a vortex dipole at depth d below the undisturbed position of the free surface. When the effects of surface tension are weak, the cusp forms at depth $2d/3$. [From Jeong & Moffatt 1992.]

3 Cusp Singularities at a Free Surface

If a viscous fluid has a free surface, and an internally-driven flow leads to strong convergence of the flow on this free surface towards a line (or curve) on it, then there is a tendency to form an inward-dipping cusp-type singularity located along this line or curve. The prototype flow here is that studied by Jeong & Moffatt (1992), in which the flow is induced by a vortex dipole of strength α placed at depth d below the undisturbed free-surface level (figure 4). The exact solution for the Stokes flow for this geometry, and with surface tension γ at the free surface, exhibits a remarkable property: the radius of curvature R of the free surface at the plane of symmetry is given by

$$\frac{R}{d} = \frac{256}{3} \exp(-32\pi Ca),\tag{6}$$

where $Ca = \mu\alpha/d^2\gamma$ is the capillary number of the flow. If we adopt the 'level-playing-field' assumption $Ca = 1$ (i.e. viscous and surface-tension effects are *a priori* given equal weight) then (6) gives $R/d \sim 10^{-42}$ (!), so that R is many orders of magnitude below the scale at which the continuum approximation is valid. From a mathematical point of view, the solution is unimpeachable; from a physical point of view (as recognised by Jeong & Moffatt 1992), it is of course completely untenable.

One possible resolution of the (physical) singularity has, in this case, been found by Eggers (2001), again through consideration of pressure effects – this time, pressure in the thin layer of air that is drawn into the cusp region (figure 5). This pressure can be determined in the first instance through the lubrication (thin-film) approximation, and the resulting modification of the viscous flow near the cusp can then be calculated. Eggers shows that the pressure field induces a runaway effect, in which air is drawn in a thin layer down into the interior of the viscous liquid (with the possibility of subsequent break-up and entrainment of bubbles into the fluid interior).

Fig. 5 Resolution of the
cusp singularity: air is
drawn into the long thin
region in the immediate
neighbourhood of the cusp,
and the air pressure gra-
dient causes deformation
and instability of the free
surface, leading to engulf-
ment of air bubbles into the
viscous fluid.

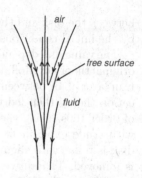

This mechanism, whereby one fluid may be drawn into the interior of another more viscous fluid, is of great potential practical importance in relation to mixing processes in chemical engineering, and merits further systematic study, both experimental and theoretical. It is noteworthy that the change of surface topology that is a characteristic feature of the mixing of two fluids (e.g. oil and vinegar) can be initiated only through the forma-tion of surface singularities, of which the cusp appears to be the naturally occurring prototype.

4 A Simple Finite-Time Singularity: the Euler Disk

Euler's disk is a toy (Bendik 2000) which exhibits a finite-time singularity in spectacular manner. It is a heavy polished steel disc which can be set in rolling motion on its slightly bevelled edge. Classical rigid-body dynamics shows that, if the point of rolling contact P describes a circle with constant angular velocity Ω, then

$$\Omega^2 \sin \alpha = 4g/a, \tag{7}$$

where a is the radius of the disk, and α the angle between its plane and the horizontal surface on which it is assumed to roll.

Dissipative effects (e.g. vibration of the supporting structure, rolling fric-tion, viscosity in the surrounding air, . . .) induce a slow decrease of α towards zero. If

$$\dot{\alpha}/\alpha \ll \Omega, \tag{8}$$

then the 'balance condition' (7) persists as an adiabatic constraint, and Ω tends to infinity as α tends to zero.

The energy (potential plus kinetic) of the disk is given by $E = \frac{3}{2}Mga \sin \alpha$, where M is the mass of the disk. The rate of dissipation of energy Φ can (in principle) be calculated as a function of E, and the equation $dE/dt = -\Phi(E)$ may then be integrated. If $\Phi(E) \sim E^\lambda$ where $\lambda < 1$, then $E \sim (t^* - t)^\beta$, where $\beta = (1 - \lambda)^{-1}$. Then from (7), $\Omega \sim (t^* - t)^{-\beta/2}$, indicating a singularity of Ω (and with it, a singularity of the vorticity at every point in the thin layer between the disk and the table) as t tends to t^*, a time determined implicitly by the initial conditions.

The value of λ depends on which dissipative mechanism is dominant. In the original theory of Moffatt (2000b), only viscous dissipation in the (ultimately) thin layer of air between the disk and the table was taken into account. Lubrication theory then led to a value $\lambda = -2$ (so $\beta = 1/3$), and gave a time t^* of order 100s (using the known parameters of the toy Euler disk) consistent with crude experiment. I have frequently been asked what happens if the disk is rolled in a vacuum, so that air viscosity as a dissipative mechanism is removed. The answer is that, unless the vacuum is extreme (so that the mean-free-path of air molecules is of the same order as the gap width αa), there is little change in the conclusion; this is because the viscosity μ of air is fairly insensitive to reduction of pressure (the mean-free-path goes up as the collision rate of molecules goes down). This is not to say that other dissipative mechanisms (notably solid rolling friction) are unimportant; however, air viscosity is the one mechanism for which a semi-quantitative description has as yet been provided.

An improved description of the influence of air viscosity, that takes account of the formation of Stokes layers on the disk and on the table, has been given by Bildsten (2002). This gives $\lambda = -5/4$, $\beta = 4/9$. The Stokes layers eventually overlap, at which stage the lubrication theory referred to above becomes applicable.

We must now ask how the singularity of Ω is to be resolved. This may be approached in two different ways, each leading to the same conclusion. If $\alpha = k(t^* - t)^\beta$, where $0 < \beta < 1$, then using (7), the adiabatic approximation (8) breaks down when $(t^* - t)^{\beta-2} \sim g/ka$. In the lubrication approximation of Moffatt (2000b), this gave $(t^* - t) \sim 10^{-2}$s, so that the singularity is averted in (literally) the last split second!

The second approach is more physical. The normal reaction at P is $N = M(g + a\ddot{\alpha})$, and the rolling condition of course requires that $N > 0$. With the above time-dependence of α, N goes to zero when $(t^* - t)^{\beta-2} \sim g/ka$, just when the adiabatic approximation breaks down. It is interesting to note that this coincidence holds independently of the dissipative mechanism (which only serves to determine the precise value of β). Thus, it would appear that in this final split second, the rolling condition (on which (7) is based) is no longer satisfied: either there is slipping at the point P, or (more probably) there is lift-off (momentary loss of contact between disk and table). A different dynamical regime is then applicable just before the disk comes finally to rest.

5 Finite-Time Singularities at Interior Points

The question of regularity of solutions of the Navier-Stokes equations continues to attract intense interest: do solutions remain smooth for all time?

Alternatively, are there any initial conditions of finite energy for which the solution exhibits a singularity at finite time? Such a singularity may conceivably occur through intense vortex stretching. But then one might normally expect the scale of the vortex core to decrease to the point at which viscous diffusion places a brake on the intensification process.

Surprisingly however, it is not necessarily the case that viscosity inevitably wins over vortex stretching, as the following simple example (from Moffatt 2000a) demonstrates. Consider the action of the time-dependent uniform strain field $\mathbf{U}$, whose components in cylindrical polar coordinates (r, θ, z) are

$$\mathbf{U} = \left(-\tfrac{1}{2}\gamma(t)r, \; 0, \; \gamma(t)z\right), \tag{9}$$

on a diffusing vortex tube in which the vorticity is

$$\boldsymbol{\omega} = (0, 0, \omega(r, t)). \tag{10}$$

The strain field $\mathbf{U}$ has infinite energy since the integral of $\mathbf{U}^2$ obviously diverges at infinity. Nevertheless, we may usefully consider its local effect (near $r = 0, z = 0$). The vorticity equation reduces to

$$\frac{\partial \omega}{\partial t} = \frac{\gamma(t)}{2r}\frac{\partial}{\partial r}(r^2\omega) + \frac{\nu}{r}\frac{\partial}{\partial r}\left(r\frac{\partial \omega}{\partial r}\right), \tag{11}$$

and we note that if γ is steady, and equal to γ_s say, then we have the familiar steady solution known as the Burgers (1948) vortex:

$$\omega = \frac{\Gamma}{\pi \delta_0^2} \exp\left(\frac{-r^2}{\delta_0^2}\right), \tag{12}$$

where $\delta_0 = 2(\nu/\gamma_s)^{1/2}$ is the 'core radius'.

Suppose now that, through some as yet unspecified mechanism, $\gamma(t)$ is given by

$$\gamma(t) = c(t^* - t)^{-1}, \quad (c > 0), \tag{13}$$

and that $\omega(r, 0)$ is given by the Gaussian formula (12). Then it is easily verified that the unique solution of (11) is

$$\omega(r, t) = \frac{\Gamma}{\nu(t^* - t)} f(\eta), \quad \eta = \frac{r}{(\nu(t^* - t))^{1/2}}, \tag{14}$$

and

$$f(\eta) = \frac{c - 1}{4\pi} e^{-\frac{1}{4}(c - 1)\eta^2}, \tag{15}$$

If $c > 1$ and $0 < t < t^*$, then this describes an approach to a singularity of vorticity on $r = 0$ as $t \to t^*$. If $0 < c < 1$, and $t^* < 0 < t - t^*$, then it describes the more familiar process of diffusion of a vortex starting from a singularity at $t = t^*$, this diffusion being attenuated by the action of the strain field.

The situation when $c = 1$ is peculiar. For $\nu > 0$, (15) gives $f(\eta) = 0$. However, if we go to the inviscid limit $\nu \to 0$, then with $t^* > 0$ and taking $c - 1 = 4\nu t^*/\delta_0^2$, (i.e. $c \to 1$ in the limit), then we have the limiting solution

$$\omega(r, t) = \frac{t^*}{t^* - t} \frac{\Gamma}{\pi \delta_0^2} \exp\left(-\frac{r^2}{\delta_0^2} \frac{t^*}{t^* - t}\right), \tag{16}$$

exhibiting singular behaviour as $t \to t^*$.

The self-similar form (14) is a special case of the Leray (1934) transformation of the Navier-Stokes equation, which may be written

$$\mathbf{u}(\mathbf{x}, t) = \left(\frac{\Gamma}{t^* - t}\right)^{\frac{1}{2}} \mathbf{U}(\mathbf{X}), \qquad \mathbf{X} = \mathbf{x}(\Gamma(t^* - t))^{-\frac{1}{2}}, \tag{17}$$

where Γ is a constant with the dimensions of circulation. Under this transformation, the vorticity takes the form

$$\boldsymbol{\omega} = \nabla \wedge \mathbf{u} = \frac{1}{(t^* - t)} \boldsymbol{\Omega}(\mathbf{X}), \qquad \boldsymbol{\Omega} = \nabla_{\mathbf{X}} \wedge \mathbf{U}, \tag{18}$$

and the vorticity equation reduces to

$$-\nabla \wedge (\mathbf{U} + \tfrac{1}{2}\mathbf{X}) \wedge \boldsymbol{\Omega} = \epsilon \nabla^2 \boldsymbol{\Omega}, \tag{19}$$

where ∇ now represents $\partial/\partial \mathbf{X}$, and $\epsilon = \nu/\Gamma$. We must obviously interpret this equation as describing flow in an inner region in a neighbourhood of the point $\mathbf{x} = \mathbf{0}$, with scale collapsing like $(t - t^*)^{1/2}$. An inner solution must match to an outer solution which is non-singular as $t \to t^*$. The only reasonable possibility (Moffatt 2000a) is

$$\boldsymbol{\Omega}(\mathbf{X}) \sim |\mathbf{X}|^{-2} \text{ as } |\mathbf{X}| \to \infty \tag{20}$$

$$\boldsymbol{\omega}(\mathbf{x}, t) \sim \Gamma|\mathbf{x}|^{-2} \text{ as } |\mathbf{x}| \to 0. \tag{21}$$

Now, following the approach of Nečas, Ružička & Šerák (1996), it has been rigorously shown by Tsai (1998) that, when $\epsilon > 0$, (19) has no nontrivial smooth solution satisfying an outer condition of the form (20). In counterbalance to this negative result, Pelz (1997, 2001) has provided strong numerical evidence (using vortex filament techniques) that, when $\epsilon = 0$, there are certain highly symmetric vorticity distributions which appear to collapse towards a singularity, following the Leray scaling (17). This singularity involves decrease of all scales (including the scale of vortex cores) in an inner region, like $(t^* - t)^{1/2}$. (See also Kerr 1997 for direct numerical simulation of Euler flows involving interacting vortices.)

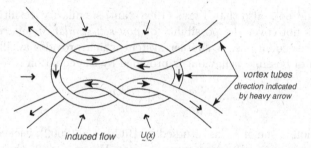

Fig. 6 A knotted vortex configuration that could conceivably give rise to a finite time singularity for the Euler equations. The vortices are conceived as inducing an inward flow contracting the knot, which must be compensated by an outward flow in the vortex tubes draining fluid from the vortices, and leading to a tightening of the knot.

Each vortex in the Pelz assemblage is subject to the strain field induced by the total vorticity distribution, and this strain increases in proportion to $(t^* - t)^{-1}$ (cf. (13)) essentially because, in the inner zone, all of the vortices approach each other, while conserving individual circulations. If this strain increases like $c(t^* - t)^{-1}$, where $c > 1$, then, as shown by the exact solution (14) of the Navier-Stokes equations, this singular behaviour should persist when ϵ is small but nonzero; the Reynolds number based on the length and velocity scales in the inner zone is then just ϵ^{-1}, which is large and which remains constant. But this description is in contradiction with Tsai's result. We escape from this contradiction only if, in the Pelz scenario, the vortices arrange themselves asymptotically in the inner zone so that the maximal strain rate acting on each vortex is $c(t - t^*)^{-1}$ with precisely $c = 1$. In this case, the singularity does not survive the transition from $\epsilon = 0$ to $\epsilon > 0$.

Figure 6 shows a hypothetical vorticity distribution (a reef knot) which might be considered as a possible candidate for an Euler singularity. Note the conical spread of vorticity (corresponding to the behaviour (20)) on leaving the 'inner zone'. The quasi-steady vorticity equation includes a term involving the outward sweeping of vorticity by the spherically symmetric 'velocity field' $\frac{1}{2}\mathbf{X}$ (with divergence 3/2); this outward sweeping must be compensated by equal and opposite inward sweeping by the induced velocity field $\mathbf{U}$, which must therefore be inwards wherever $\mathbf{\Omega}$ has a nonzero nonradial component, and which, being divergenceless, must be outwards only where $\mathbf{\Omega}$ is exactly radial. It is difficult to see how the transition from inward to outward flow can be compatible with these requirements.

Similar considerations must apply to the Pelz configuration (or to any Euler flow exhibiting collapse to a singularity with Leray scaling): inflow across the vorticity field must be compensated by outflow only in regions where the vorticity is radial. It remains to be shown whether this can actually be achieved.

We should note also that Tsai's (1998) 'anti-singularity' result, described above, does not cover the possibility of *non-self-similar singularities* of the Navier-Stokes equations. A simpler system which provides an illuminating illustration of possible behaviour is provided by the equation

$$\frac{\partial \phi}{\partial t} = \phi^2 + \epsilon \frac{\partial^2 \phi}{\partial t^2} \,. \tag{22}$$

This equation is one of a class studied by Budd et al (2002); it clearly admits self-similar singular behaviour when $\epsilon = 0$. When $\epsilon \neq 0$, the singularity survives (at a shifted singularity time), but it is no longer self-similar in character.

References

1. Bendik, J. (2000) Website: www.eulerdisk.com
2. Bildsten, L. (2002) Dissipation for Euler's disk and a desktop demonstration of coalescing neutron stars. *Phys. Rev. E* **66**, 056309-10
3. Budd, C.J., Collins, G.J. & Galaktionov, V. , (1998) Asymptotic and numerical description of self-similar blow-up in quasilinear parabolic equations. *J. Comp. Appl. Math.* **97**, 51–80.
4. Burgers, J.M. (1948) A mathematical model illustrating the theory of turbulence. *Adv. Appl. Mech.* **1**, 171–199.
5. Eggers, J. (2001) Air entrainment through free-surface cusps. *Phys.Rev.Letters* **86**, 4290–4293.
6. Jeong, J-T. & Moffatt, H.K. (1992) Free-surface cusps associated with flow at low Reynolds number. *J. Fluid Mech.* **241**, 1–22.
7. Kerr, R.M. (1997) Euler singularities and turbulence. In *Proc.19th ICTAM* (Eds. T. Tatsumi, E. Watanabe & T. Kambe), Elsevier Science Publ., Amsterdam, 57–70.
8. Leray, J. (1934) Sur le mouvement d'un liquide visqueux emplissant l'espace. *Acta Math.* **63**, 193–248.
9. Moffatt, H.K. (2000a) The interaction of skewed vortex pairs: a model for blow-up of the Navier-Stokes equations. *J. Fluid Mech.* **409**, 51–68.
10. Moffatt, H.K. (2000b) Euler's disk and its finite-time singularity. *Nature* **404**, 833–834.
11. Moffatt, H.K. (2001) Local and global perspectives in fluid dynamics. In *Mechanics for a New Millennium* (Eds. H. Aref & J. W. Philips), Kluwer Academic Publ. 521–540.
12. Nečas, J., Ružička, M. & Šerák (1996) On Leray's self-similar solutions of the Navier-Stokes equations. *Acta Math.* **176**, 283–294.
13. Pelz, R.B. (1997) Locally self-similar finite-time collapse in a high-symmetry vortex filament model. *Phys. Rev. E* **55**, 1617–1626.
14. Pelz, R. B. (2001) Symmetry and the hydrodynamic blow-up problem. *J.Fluid Mech.* **444**, 299–320.
15. Taylor, G.I. (1960) Similarity solutions of hydrodynamic problems. In *Aeronautics and Astronautics* (Durand Anniversary Vol), Pergamon, 21–28.
16. Tsai T-P. (1998) On Leray's self-similar solutions of the Navier-Stokes equations satisfying local energy estimates. *Arch. Rational Mech. Anal.* **143**, 29–51.

Structural Complexity and Dynamical Systems

Renzo L. Ricca
(School Director and CIME Lecturer)

Abstract With this paper we want to pay tribute to 150 years of work on topological fluid mechanics. For this, we review Helmholtz's (1858) original contribution on topological issues related to vortex motion. Some recent results on aspects of structural complexity analysis of fluid flows are presented and discussed, as well as new results on topological bounds on the energy of magnetic knots and links in ideal magnetohydrodynamics, and on helicity-crossing number relations in dissipative fluids.

1 Introduction

The origin of topological fluid mechanics is probably rooted in the works on vortex motion by Helmholtz (1858) and Lord Kelvin (1869), and much of its modern developments are due to the formidable recent progress in knot theory, vector field analysis, mathematical fluid dynamics and computational visualization. With this paper we want to pay tribute to these 150 years of work on topological fluid mechanics. §1 is dedicated to review Helmholtz's contribution in the light of modern developments in mathematical fluid dynamics: since much of Helmholtz's original emphasis on the relevance of topological issues in fluid mechanics has gradually disappeared from textbooks on vortex dynamics, we simply re-examine his work, emphasizing the merits he deserves for this. §2 overviews some of my recent work on aspects of structural complexity analysis of fluid flows, including a brief summary of

R.L. Ricca
Department of Mathematics and Applications, University of Milano-Bicocca,
Via Cozzi 53, 20125 Milano, ITALY.
e-mail: renzo.ricca@unimib.it
http://www.matapp.unimib.it/~ricca/
and
Institute for Scientific Interchange, Villa Gualino, 10133 Torino, ITALY.

R.L. Ricca (ed.), *Lectures on Topological Fluid Mechanics*,
Lecture Notes in Mathematics 1973, DOI: 10.1007/978-3-642-00837-5,
© Springer-Verlag Berlin Heidelberg 2009

some current work on applications of critical point theory and topology-based visiometrics. Finally, §3 presents some new results on topological bounds on the energy of magnetic knots and links in ideal magnetohydrodynamics and on helicity-crossing number relations in dissipative fluids.

2 Helmholtz's Work on Vortex Motion: Birth of Topological Fluid Mechanics

It is perhaps little known that the seminal work of Helmholtz (1858; hereafter referred to as H58) *On integrals of the hydrodynamical equations, which express vortex-motion* (Tait's translation) pioneers fundamental questions in topological fluid mechanics. In many ways this is a truly remarkable paper. In addressing and solving the problem of determining rotational motion of fluid elements, for which a single-valued velocity potential cannot be defined, Helmholtz demonstrates three conservation laws for vortex motion (see below), that have become a cornerstone in the foundation of mathematical fluid mechanics (see, for example, Saffman, 1991). The undisputed importance of his main contributions, i.e. the discovery of the conservation laws of vortex motion, has, however, gradually shadowed the strong topological flavour, that permeates the whole paper from the very start. Helmholtz's investigation moves indeed from Euler's original observation of 1755, that even in absence of a kinetic potential certain types of fluid motion are nevertheless possible. In analyzing the conditions of motion, Helmholtz establishes the existence of two classes of hydrodynamic integrals by identifying two separate domains of definition, where either a single-, or a multiple-valued velocity potential is defined, depending on the degree of connection of the fluid region. Conditions for which a single-valued velocity potential exists and the relationship with the multiplicity of connection of the ambient space are discussed in §1 of H58, where it is shown (p. 486) *"that when there is a velocity-potential the elements of the fluid have no rotation, but that there is at least a portion of the fluid elements in rotation when there is no velocity-potential."*

2.1 Multi-Valued Potentials in Multiply Connected Regions

Following Helmholtz's discussion, let us consider an ideal, incompressible fluid in an unbounded, *simply connected* domain $\mathcal{D}$ of $\mathbb{R}^3$. Motion is governed by the standard Euler's equations, supplemented by the incompressibility condition, that is

$$\frac{D\mathbf{u}}{Dt} = -\nabla p + \mathbf{f} , \qquad \mathbf{u} = 0 \quad \text{as} \quad \mathbf{X} \to \infty , \tag{1}$$

with
$$\nabla \cdot \mathbf{u} = 0 , \tag{2}$$

where $\mathbf{u} = \mathbf{u}(\mathbf{X}, t)$ is the velocity of a fluid particle at position $\mathbf{X}$ and time t, p is pressure, $\mathbf{f}$ denotes conservative forces, and fluid density is set to be equal to 1 for convenience. All functions are assumed to be sufficiently smooth at all times. In absence of rotation, we can define a velocity potential ϕ everywhere in $\mathcal{D}$, that is

$$\nabla \times \mathbf{u} = 0 \quad \rightarrow \quad \mathbf{u} = \nabla \phi ; \tag{3}$$

incompressibility, then, yields the Laplace equation for ϕ, i.e.

$$\nabla \cdot \mathbf{u} = 0 \quad \rightarrow \quad \nabla \cdot (\nabla \phi) = \nabla^2 \phi = 0 , \tag{4}$$

with
$$\phi = \text{cst.} \quad \text{as} \quad \mathbf{X} \rightarrow \infty , \tag{5}$$

which determines a harmonic, single-valued potential function everywhere in $\mathcal{D}$. Thus, the corresponding integrals of motion are said to be *integrals of the first class* (H58, p. 499). Note that the existence of a single-valued velocity potential ϕ is due to the condition $\nabla \times \mathbf{u} = 0$, everywhere in $\mathcal{D}$. In the language of differential forms this is summarized by the following fundamental relations:

Theorem (Fundamental correspondence). *Let $\mathcal{D}$ be simply connected. Then, we have*

$$\left\{ \begin{matrix} \nabla \times \mathbf{u} = 0 \\ \mathbf{u} \text{ irrotational} \end{matrix} \right\} \quad \longleftrightarrow \quad \left\{ \begin{matrix} \mathrm{d}\alpha = 0 \\ \alpha \text{ closed 1-form} \end{matrix} \right\}$$

$$\Updownarrow \qquad\qquad\qquad\qquad \Updownarrow$$

$$\mathbf{u} = \nabla \phi \qquad\qquad \longleftrightarrow \qquad\qquad \alpha = \mathrm{d}\beta$$

$$\Updownarrow \qquad\qquad\qquad\qquad \Updownarrow$$

$$\left\{ \begin{matrix} \mathbf{u} \text{ conservative} \\ \phi \text{ single-valued potential} \end{matrix} \right\} \quad \longleftrightarrow \quad \left\{ \begin{matrix} \alpha \text{ exact} \\ \beta \text{ 0-form} \end{matrix} \right\}$$

$$\Updownarrow \qquad\qquad\qquad\qquad \Updownarrow$$

$$\int_{\mathcal{C}} \mathbf{u} \cdot \mathrm{d}\mathbf{l} , \quad \mathcal{C}\text{-independent} \quad \longleftrightarrow \quad \int_{\mathcal{C}} \alpha , \quad \mathcal{C}\text{-independent}$$

$$\Updownarrow \qquad\qquad\qquad\qquad \Updownarrow$$

$$\oint_{\mathcal{C}_0} \mathbf{u} \cdot \mathrm{d}\mathbf{l} = 0 , \quad \forall \mathcal{C}_0 \text{ in } \mathcal{D} . \quad \longleftrightarrow \quad \oint_{\mathcal{C}_0} \alpha = 0 , \quad \forall \mathcal{C}_0 \text{ in } \mathcal{D} .$$

Proof. (Sketch.) Considering first the l.h.s. column, proof of the first top two relations is given by Helmholtz (H58, p. 488–490), while relations at the bottom are consequences of the application of Stokes' theorem. As regards the r.h.s. column, we can easily see that if α is a closed 1-form, then, by definition, $\mathrm{d}\alpha = 0$ and since $\mathcal{D}$ is simply connected, then it is known the de Rham cohomology group of every closed 1-form on $\mathcal{D}$ is exact (see, for example, Bott & Tu, 1982); hence $\alpha = \mathrm{d}\beta$. On the other hand if α is exact, then, by definition, $\alpha = \mathrm{d}\beta$ and $\mathrm{d}\alpha = \mathrm{d}(\mathrm{d}\beta) = 0$, since every exact differential form is closed. Then, let $\mathcal{C}$ be an oriented smooth 1-manifold in $\mathcal{D}$: α is exact on $\mathcal{C}$ if and only if $\int_{\mathcal{C}} \alpha$ is path-independent on $\mathcal{D}$. Moreover, if $\mathcal{C} = \mathcal{C}_0$ is closed, then, by corollary, $\oint_{\mathcal{C}_0} \alpha$ is path-independent if and only if for every closed 1-manifold in $\mathcal{D}$, $\oint_{\mathcal{C}_0} \alpha = 0$. For the one-to-one correspondences between the two columns see, for example, Weintraub (1997, p. 27 and p. 117).   □

If $\boldsymbol{\omega} = \nabla \times \mathbf{u} \neq 0$ in some region $\mathcal{W} \subset \mathcal{D}$ (and $\nabla \times \mathbf{u} = 0$ everywhere else in $\mathcal{D}/\mathcal{W}$), Helmholtz shows (H58, p. 489–490) that the velocity field $\mathbf{u}$ cannot be given by a single velocity potential ϕ defined in $\mathcal{D}/\mathcal{W}$ through $\nabla\phi$. The presence of a rotational region $\mathcal{W}$, embedded in an unbounded irrotational fluid, makes the irrotational region $\mathcal{D}/\mathcal{W}$ (the complement to $\mathcal{W}$ in $\mathbb{R}^3$) multiply connected. Here Helmholtz refers to the new concepts just developed by Riemann (1857) on multiply connected surfaces. A surface in $\mathbb{R}^2$ is said to be n-ply connected, if there are at most $n-1$ independent, distinct simple circuits, i.e. simple closed paths, irreducible to a point and to one another. The plane in $\mathbb{R}^2$ is an example of simply connected surface (top of Figure 1a), whereas a doubly connected surface has one hole in it (see bottom of Figure 1a), the latter representing, for example, a region of rotation. Extension of these concepts to $\mathbb{R}^3$ is straightforward (Figure 1b).

In a simply connected region every closed path (circuit) is reducible to a point, thus by Stokes' theorem the circulation is zero everywhere (Figure 2a). If the region is doubly connected, though, there is at most one *simple*, irreducible circuit (Figure 2b), whose circulation has a finite value, say κ,

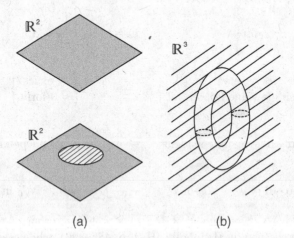

(a)   (b)

Fig. 1 (a) Simply (top) and doubly connected region in $\mathbb{R}^2$. (b) A vortex ring, defined on a toroidal domain $\mathcal{W} \subset \mathcal{D}$, embedded in an unbounded, irrotational fluid domain, is a doubly-connected region in $\mathbb{R}^3$; note that the complement $\mathcal{D}/\mathcal{W}$, filled by irrotational fluid, is also doubly-connected.

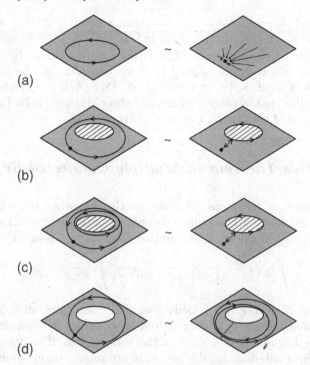

Fig. 2 (a) Every circuit drawn in a simply connected region is reducible to a point, hence the circulation is zero. (b) In a doubly-connected region there is at most one irreducible simple circuit, whose circulation has a finite value, say κ. (c) Example of a multiple circuit ($m = 2$) in a doubly connected region: in this case the circulation is 2κ. (d) A doubly connected region is reduced to a simply connected one by cutting the region with a "circulation-stopping" barrier, represented in figure by the black line.

which is the *cyclic constant* of the region. The circuit is not simple but *multiple*, if it encircles the rotational region m times, as in Figure 2c; the region's cyclic constant is then $m\kappa$ and in the case of an n-ply connected region the *cyclosis* of the region is given by $\sum_{i=1}^{n-1} m_i \kappa_i$. An n-ply connected region can be reduced to a simply connected one by inserting $n-1$ cuts given by $n-1$ separatrix surfaces drawn across the region (see the case of Figure 2d), each cutting surface diminishing the degree of connectedness by one. These $n-1$ separatrices act as "stopping barriers" (adopting Lord Kelvin's terminology) to the circulation around the rotational regions, each insertion contributing to the total bounding surface of $\mathcal{D}$. For fluid motions in multiply connected regions the velocity potential takes indeed more than one value. Since the velocity is proportional to the differential coefficients of ϕ, fluid flows are given by ever increasing values of ϕ. But for a fluid particle that moves on a path encircling a rotational region $\mathcal{W}$, as the particle returns to its original position, the potential ϕ attains a second greater value. Hence (H58, p. 499), *"since this may occur indefinitely, there must be for every point of* *such a complexly-connected space an infinite number of distinct values of ϕ differing by equal quantities like those of* $\tan^{-1}(x/y)$, *which is such a many-valued function and satisfies the differential equation"* given by the second of (4). These observations, applied to fluid dynamics, will be investigated further, and in great depth, by Lord Kelvin (1868, Art. 54 to end), and subsequently elaborated by other authors, including Maxwell (1873, Preliminary, Art. 18–22) and Lamb (1879, Chapter 3, Art. 47–55).

2.2 Green's Theorem in Multiply Connected Regions

Helmholtz (H58, p. 488, footnote) makes another important remark regarding the inapplicability of Green's (first) theorem in presence of rotational motion and multi-valued functions. The standard theorem by Green states that

$$\int_V [\phi \nabla^2 \phi' + (\nabla \phi) \cdot (\nabla \phi')] \, \mathrm{dV} = \int_S (\phi \nabla \phi') \cdot \mathrm{dS} \, , \tag{6}$$

where ϕ and ϕ' denote two velocity potentials, and integration is intended over the volume of the fluid domain and its bounding surface, dS denoting an outward-drawn vector element of the surface area. If rotational motion is present in a sub-domain, the region is no longer simply connected and the indeterminacy associated with velocity potentials (assuming that their gradients are single-valued) invalidates the theorem. By taking account of the separatrix surfaces inserted to make the region simply connected, Kelvin (1869) amends the theorem as follows

$$\int_V [\phi \nabla^2 \phi' + (\nabla \phi) \cdot (\nabla \phi')] \, \mathrm{dV}$$
$$= \int_S (\phi \nabla \phi') \cdot \mathrm{dS} + \sum_{i=1}^{n-1} m_i \kappa_i \int_{\Sigma_i} \nabla \phi' \cdot \mathrm{d}\Sigma_i \, , \tag{7}$$

where integration in the r.h.s. of eq. (7) is now augmented by the sum of the integrals extended to the $n-1$ separatrices of area Σ_i $(i = 1, \ldots, n-1)$, each κ_i denoting the jump in ϕ (hence in circulation) across Σ_i. Kelvin's extension of Green's theorem to multiply connected domains may find useful applications in current theoretical physics: from topological quantum field theory to cosmological models, in presence of black-holes and topological defects.

2.3 Conservation Laws

By considering rotation confined to a tubular-like region (that is a *vortex filament*) embedded in an irrotational fluid, Helmholtz proceeds to prove three laws of conservation for vortex motion. In his own words (H58, §2), he states that:

Theorem (Helmholtz's conservation laws).

1. *Elements of the fluid which at any instant have no rotation, remain during the whole motion without rotation.*
2. *Each vortex line remains continually composed of the same elements of fluid.*
3. *The product of the section and the angular velocity, in a portion of a vortex filament containing the same element of fluid, remains constant during the motion of that element.*

Note the topological character of the first two statements, that can be seen as complement to one another. In modern terms we simply say that a region of vorticity $\mathcal{W}$, embedded in an unbounded, irrotational fluid $\mathcal{D}/\mathcal{W}$, is *frozen* in the fluid and is isotoped to the new region $\varphi_t(\mathcal{W})$ by diffeomorphisms of the flow map φ, by preserving its rotational character at any time; similarly so for the irrotational fluid in the complement region.

3 Measures of Structural Complexity

In recent years the demand for advanced diagnostic tools for computational vortex dynamics, turbulent flows and magnetohydrodynamics has grown considerably (see, for example, Weickert & Hagen, 2006). Detailed analysis of space localization and time evolution of coherent structures, defined by statistical coherence of physically relevant quantities — be these passive scalars, vector or tensor fields — requires new tools to quantify structural complexity present in the fluid (Ricca, 2000; 2001; 2005). Mathematical concepts borrowed from differential geometry, knot theory, graph theory, dynamical systems theory and other branches of modern mathematics can be usefully employed in numerical analysis of direct numerical simulations of fluid flows to quantify, estimate or infer production, transfer and depletion of physical quantities such as energy and momentum. Current research is mainly oriented in the following directions:

Theoretical goals:

i)   to describe and classify complex morphologies;
ii)   to study relationships between complexity and energy;
iii) to understand and predict energy localization and transfer.

Applications:

i)   to implement new visiometric tools and diagnostics;
ii)   to develop real-time energy analysis of dynamical processes;
iii) to compare estimated values with expected values of standard models.

3.1 Dynamical Systems and Vector Field Analysis

A lot of work has been done in this direction, and most notably on the implementation of structural classification of vector fields, on structural stability analysis and on visualization and processing of tensor fields. A very brief summary and a few references are given here for convenience.

Structural classification of vector fields
Structural classification of three-dimensional vector fields $\mathbf{v}(\mathbf{X})$ relies mainly on the eigenvalue/eigenvector analysis of the Jacobian matrix $\mathbf{J_v}(\mathbf{X}) = \nabla\mathbf{v}(\mathbf{X})$ (see, for example, Chong *et al.*, 1990). A first-order critical point $\mathbf{X}_0$, given by the condition $\mathbf{v}(\mathbf{X}_0) = 0$, can be classified according to the order and value of the real parts of the eigenvalues of $\mathbf{J_v}(\mathbf{X}_0)$, provided $\det(\mathbf{J_v}(\mathbf{X}_0)) \neq 0$. Let $\Re e(\lambda_1) \leq \Re e(\lambda_2) \leq \Re e(\lambda_3)$ be the ordered real parts of the eigenvalues; critical points can be classified according to the following scheme:

(i) source : $0 \quad < \Re e(\lambda_1) \leq \Re e(\lambda_2) \leq \Re e(\lambda_3)$

(ii) repelling saddle : $\Re e(\lambda_1) < \quad 0 \quad < \Re e(\lambda_2) \leq \Re e(\lambda_3)$

(iii) attracting saddle : $\Re e(\lambda_1) \leq \Re e(\lambda_2) < \quad 0 \quad < \Re e(\lambda_3)$

(iv) sink : $\Re e(\lambda_1) \leq \Re e(\lambda_2) \leq \Re e(\lambda_3) < \quad 0$

where outflow/inflow direction is given by the sign of $\Re e(\lambda_i)$: a negative real part implies inflow (attracting direction) and positive real part implies outflow (repelling direction). Each critical point can be further classified in two families:

(a) focus : $\Im m(\lambda_i) = 0 \quad \text{and} \quad \Im m(\lambda_j) = -\Im m(\lambda_k) \neq 0$

(b) node : $\Im m(\lambda_i) = \Im m(\lambda_j) = \Im m(\lambda_k) = 0$

where $\Im m(\cdot) =$ denotes imaginary part and $i \neq j \neq k$, $\{i, j, k\} \in \{1, 2, 3\}$, the imaginary part implying circulation. Figure 3 shows an early example of structural classification of streamlines in a three-dimensional separation flow.

Structural stability of dynamical systems
Structural stability issues of dynamical systems rely greatly on the results by Morse, Smale and Peixoto and there is now a wealth of information on divergence-free fields on two-dimensional compact manifolds, motivated by applications to geophysical fluid dynamics. The interested reader may refer to the book by Ma & Wang (2005) for latest results and some generalization to three dimensions.

Visualization and processing of tensor fields
From the early 1990s geometry and topology-driven visualizations have been steadily developed from progress made on structural complexity analysis and

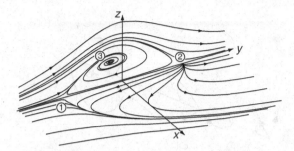

Fig. 3 Study of a three-dimensional separation flow by analysis of the surface streamline pattern (on the $x - y$ plane) and of the solution trajectories (on the plane of symmetry $y - z$). Note that there are 3 major critical points: point 1 is a no-slip saddle, point 2 is a no-slip node in the $x - y$ plane, and point 3 is a free-slip focus in the $y - z$ plane (from Chong *et al.*, 1990).

critical point theory. These methods are of increasing importance in the analysis and visualization of data-sets from a wide variety of scientific domains. Current challenges include the management of time-dependent data, feature extraction and representation of large and complex data-sets, multi-scale adaptive visualization. The interested reader may consult the collection of papers edited by Hauser *et al.* (2007).

3.2 Measures of Tangle Complexity

Computational fluid dynamics (CFD) produces data-sets of numerical simulations, from which we can extract numerical sub-domains, representing for example vortical or turbulent regions, magnetic field or passive scalar distributions, to analyze. Preliminary steps to any study include: (i) identifying a prescribed sub-domain of physical interest (defined as the *tropicity domain*) and the corresponding characteristic scales, (ii) determining the characteristic dimensions of the chosen region, (iii) assigning a reference system. The physical problem and numerical threshold associated with the CFD code will give information on step (i), while step (ii) and (iii) serve to perform analysis on structural complexity.

Tropicity dimensions and tropicity directions
Let $\mathcal{T} = \bigcup_i \chi_i$ $(i = 1, \ldots, n)$ be the n-component tangle given by the set of vector fields (such as streamlines, vortex lines or magnetic fields) or solution trajectories (e.g. pressure or temperature distributions) to be analyzed. For each tangle component χ_i we can determine the maximal *tropicity dimensions,* given by

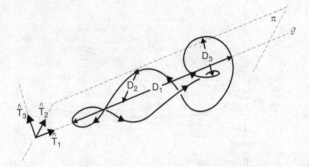

Fig. 4 Tropicity dimensions and reference basis determined by the tangle component χ_i.

$$
\left.
\begin{aligned}
D_1^{(i)} &= \overline{P_0 P_1} \equiv \max_{\{j,k\}} d(P_j, P_k) \,, \\
D_2^{(i)} &= \overline{OP_2} \equiv \max_j d(P_j, \ell(P_0, P_1)) \,, \\
D_3^{(i)} &= \overline{QP_3} \equiv \max_j d(P_j, \pi(P_0, P_1, P_2)) \,,
\end{aligned}
\right\} \tag{8}
$$

where the points P_j, P_k are sampled over χ_i; hence, the principal *tropicity unit vectors* are given by

$$
\left.
\begin{aligned}
\hat{\mathbf{T}}_1^{(i)} &\equiv (P_1 - P_0)/D_1^{(i)} \,, \\
\hat{\mathbf{T}}_2^{(i)} &\equiv (P_2 - O)/D_2^{(i)} \,, \\
\hat{\mathbf{T}}_3^{(i)} &\equiv \hat{\mathbf{T}}_1^{(i)} \times \hat{\mathbf{T}}_2^{(i)} \,.
\end{aligned}
\right\} \tag{9}
$$

$\{\hat{\mathbf{T}}_1^{(i)}, \hat{\mathbf{T}}_2^{(i)}, \hat{\mathbf{T}}_3^{(i)}\}$ define the reference system on χ_i, (see Figure 4). By averaging this information over n components, we obtain the tropicity dimensions and the reference system of the tangle, that is

$$
\left.
\begin{aligned}
D_1 &= \langle D_1^{(i)} \rangle \,, & \hat{\mathbf{T}}_1 &= \langle \hat{\mathbf{T}}_1^{(i)} \rangle \,, \\
D_2 &= \langle D_2^{(i)} \rangle \,, & \hat{\mathbf{T}}_2 &= \langle \hat{\mathbf{T}}_2^{(i)} \rangle \,, \\
D_3 &= \langle D_3^{(i)} \rangle \,, & \hat{\mathbf{T}}_3 &= \langle \hat{\mathbf{T}}_3^{(i)} \rangle \,,
\end{aligned}
\right\} \tag{10}
$$

Note that the tangle tropicity vectors, are determined by the global geometry of the tangle, and in general do not coincide with the eigenvectors of the Jacobian matrix of the velocity gradients.

Tangle analysis by indented projections
Tropicity directions find applications in tangle analysis. The latter is based on the concept of projected diagrams. Let us consider first a single component χ (for simplicity we shall drop the suffix) and its "indented" projection χ_p, obtained by the orthogonal projection p on the plane of projection Π_p (see Figure 5): by allowing small indentations at crossing sites of the projected

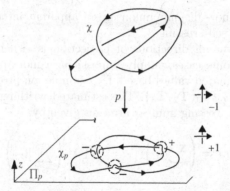

Fig. 5 Indented projection of the tangle component χ under p. The signs at each crossing site of the indented diagram are assigned according to the convention rule shown on the right-hand-side.

curve, the indented projection retains the information associated with over- and under-passes of the original curve viewed along the projection direction. Evidently χ_p depends on the direction of projection, and any change in the latter is obviously reflected in the shape of χ_p. Topological information can be recovered by implementing the three Reidemeister moves computationally, to reduce χ_p to its minimal form, i.e. with minimal number of crossings. By assigning the value $\epsilon_r = \pm 1$ to each apparent crossing of χ_p, according to the sign convention shown in Figure 5, we can compute three important quantities. The first is the *writhing number* Wr, given by

$$Wr_i = Wr(\chi_i) = \left\langle \sum_{r \in \chi_i} \epsilon_r \right\rangle , \qquad Wr = Wr(\mathcal{T}) = \left\langle \sum_{r \in \mathcal{T}} \epsilon_r \right\rangle , \qquad (11)$$

where here brackets denote averaging over all directions of projection. This quantity gives geometric information on average chirality and associated degree of three-dimensional average coiling in the tangle.

A second important quantity is the total *linking number* Lk_{tot}, given by

$$Lk_{ij} = Lk(\chi_i, \chi_j) = \frac{1}{2} \sum_{\substack{r \in \chi_i \cap \chi_j \\ i \neq j}} \epsilon_r , \qquad Lk_{\text{tot}} = \sum_{i \neq j} |Lk_{ij}| , \qquad (12)$$

where $\sqcap$ denotes disjoint union on the apparent intersections of curve strands, omitting self-crossings. This quantity provides topological information on tangle complexity and changes with the recombination of tangle components.

A third quantity, of algebraic character, that provides a good measure of structural complexity is given by the *average crossing number* $\overline{C}$, defined by the sum of all un-signed crossings, averaged over all projections. We have

$$\overline{C}_{ij} = \overline{C}(\chi_i, \chi_j) = \left\langle \sum_{r \in \chi_i \# \chi_j} |\epsilon_r| \right\rangle , \qquad \overline{C} = \sum_{r \in \mathcal{T}} \overline{C}_{ij} , \qquad (13)$$

where $\#$ denotes now disjoint union on all apparent intersections of curve strands, including self-crossings.

The average over all directions of projection is either computationally expensive or, in some cases, simply impossible: a more practical approach is to resort to estimated values based, for example, on projections along the tropicity directions $\{\hat{\mathbf{T}}_1, \hat{\mathbf{T}}_2, \hat{\mathbf{T}}_3\}$. The estimated writhing number and the estimated average crossing number are thus given by

$$Wr_\perp = \left(\sum_{r\in\mathcal{T}} \epsilon_r \right)_\perp , \qquad \overline{C}_\perp = \left(\sum_{r\in\mathcal{T}} |\epsilon_r| \right)_\perp , \qquad (14)$$

where $\perp$ denotes the algebraic mean over the three principal projections. Additional information comes also from classical geometric and topological analysis on tangle components (see, for instance, Ricca, 2000).

Comparative analysis on a test case: superfluid vortex tangle

Comparative analysis on the above measures has been conducted by direct numerical simulations of vortex tangles (Barenghi et al., 2001; 2002), produced by perturbations due to a background flow field, or by a decaying turbulent field. Here we report the test case of a superfluid vortex tangle, produced by a background ABC-type of flow. Complexity measures are extracted from the data-sets as the tangle grows in time (see Figure 6) and analysis of growth rates is performed when the tangle is fully developed (between $t = 0.06$ and $t = 0.09$). Physical quantities such as kinetic helicity H, nor-

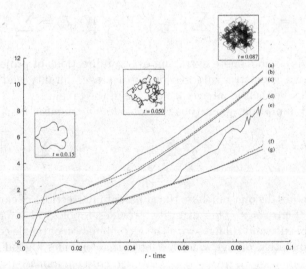

Fig. 6 Comparative analysis of complexity measures for superfluid vortex tangle simulation: tangle mature growth is shown in inset at $t = 0.087$. (a) $\log|H|$; (b) $\log\overline{C}_\perp$; (c) $\log\overline{C}$; (d) $\log Wr_\perp$; (e) $\log Lk_{\mathrm{tot}}$; (f) $\log(E/E_0)$; (g) $\log(L/L_0)$ (from Barenghi et al., 2001).

malized kinetic energy E/E_0 and normalized total length L/L_0 are also shown for comparison and check.

The mature stage is characterized by two distinct growth rates: energy/length growth rate $\approx O(83s^{-1})$; complexity measures and helicity growth rate $\approx O(165s^{-1})$. It is remarkable that essentially all the measures tested have similar growth rate. In any case the average crossing number seems to be the most appropriate candidate measure of structural complexity. Comparison between the theoretical value $\overline{C}$, obtained by implementing the analytical definition by Freedman & He (1991) and its estimated value $\overline{C}_\perp$, shows that the estimated value — much more convenient computationally — approximates very well the theoretical value. Note also that the discontinuous behaviour of Lk_{tot} is given by topological changes associated with vortex reconnection, but its mean growth rate does not differ significantly from that of other measures.

Tangle analysis by signed area information

If we consider standard projection, instead of indented projection, the projected diagram is a planar nodal curve. In case of thin vortex filaments, as in superfluids, signed area information extracted from the planar graph may be used to estimate linear and angular momentum associated with the vortex tangle. This is based on the interpretation of momenta in terms of projected area and on the application of a geometric method to calculate the signed area of complex graphs (for details, see Ricca, 2008b). Here we want to illustrate briefly this method.

The oriented graph diagram of a tangle of vortex lines is an oriented nodal curve in $I\!R^2$, and this often attains considerable complexity, particularly as regards the localization of its self-intersections. A necessary first step is to reduce nodal curves of any complexity to *good* nodal curves, that have (at most) double points. Nodal points are classified according to their degree of multiplicity $\mu(P)$ given by the number of arcs incident at the point of intersection P. If P is a double point, then $\mu(P) = 2$. If P is a point of multiplicity $\mu(P) = n$ ($n > 2$), we can always reduce its multiplicity by "shaking" the graph diagram (actually its pre-image) near P to get $m = \frac{1}{2}(n^2 - n)$ double points, by virtual perturbations of the incident arcs from their location. Thus, if $h^{(n)}$ is the total number of points of multiplicity n, by applying this shaking technique we can always replace these $h^{(n)}$ points with $h(n) = mh^{(n)}$ ($m \geq 3$) double points. We say that a graph diagram is a *good projection*, when it has at most double points. Hence, by the shaking technique, we can always reduce highly complex graph diagrams to good nodal curves.

Let $\mathcal{C}$ denote one of such good nodal curves on Π, and let $A(\mathcal{C})$ be the corresponding total area. In order to calculate this area, first we need to define the index $\mathcal{I}_P(\mathcal{C})$ of $\mathcal{C}$ associated with any internal point P. Let $P \notin \mathcal{C}$, $\hat{t}$ the tangent to $\mathcal{C}$ and ρ the radiant vector with foot at P, that intersects $\mathcal{C}$ transversally. At each intersection point $X \in \rho \cap \mathcal{C}$ assign the algebraic sign $\epsilon(X) = \pm 1$, according to the standard convention given by the right-hand rule, that is $\epsilon(X) = +1$ when the frame $\{\rho, \hat{t}\}$ is positive (see Figure 7a).

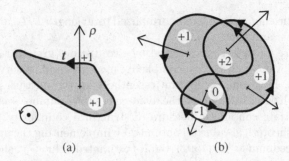

(a)   (b)

Fig. 7 (a) The number in the dashed region is the value of the index $\mathcal{I}_P(\mathcal{C})$ according to the right-hand rule convention and the algebraic intersection number calculated by eq. (15). (b) The oriented nodal curve, resulting, for example, from the standard projection of a figure-8 knot, has 5 bounded regions. Note that one of the interior regions has index 0, due to the opposite orientation of the strands crossed by ρ.

If X is a double point, then the intersection is computed with one of the neighbouring pairs of the incident, equi-oriented arcs.

Definition. The *index* $\mathcal{I}_P(\mathcal{C})$ of $\mathcal{C}$ at P is the algebraic intersection number given by

$$\mathcal{I}_P(\mathcal{C}) = \sum_{X \in \rho \cap \mathcal{C}} \epsilon(X). \tag{15}$$

Let us now consider the Z sub-domains $\{\mathcal{R}_j\}_{j=1,\dots,Z}$ determined by $\mathcal{C} \cap \Pi$ and bounded by $\mathcal{C}$, and let $A(\mathcal{R}_j) > 0$ denote their standard area. Since every point $P \in \mathcal{R}_j$ has the same $\mathcal{I}_P(\mathcal{C})$, we shall call $\mathcal{I}_j$ the index associated with any point $P \in \mathcal{R}_j$ and assign this value to each sub-domain $\mathcal{R}_j$ of $\mathcal{C} \cap \Pi$ (see Figure 7b). The signed area of an oriented graph, a concept that can be traced back to Gauss, is thus given by the following rule.

Rule (Signed area). *The signed area $A(\mathcal{C})$ of an oriented, planar nodal curve $\mathcal{C}$, is given by*

$$A(\mathcal{C}) = \sum_{j=1}^{Z} \mathcal{I}_j A(\mathcal{R}_j) \tag{16}$$

where $A(\mathcal{R}_j) > 0$ is the standard area of $\mathcal{R}_j$.

By the signed area rule we can calculate the projected area of any nodal curve, be it the graph of a single vortex line, or that of a complex tangle of vortices. If the vortices have different circulations, a weighting factor defined in terms of contributions from each arc of $\partial \mathcal{R}_j$ must be assigned to $A(\mathcal{R}_j)$. One of the simplest correction comes from an algebraic weighting γ_j of the circulations associated with $\partial \mathcal{R}_j$. Thus, for thin vortices evolving under the so-called localized induction approximation (LIA, for short), we can prove (Ricca, 2008b) the following result:

Theorem (Signed area interpretation). *Let $\mathcal{T}$ be a vortex tangle evolving under LIA. Then, the linear momentum $\mathbf{P} = \mathbf{P}(\mathcal{T})$ has components*

$$P_{xy} = \sum_{j=1}^{Z} \gamma_j \mathcal{I}_j A_{xy}(\mathcal{R}_j), \; P_{yz} = \ldots, \; P_{zx} = \ldots, \tag{17}$$

and the angular momentum $\mathbf{M} = \mathbf{M}(\mathcal{T})$ has components

$$M_{xy} = d_z \sum_{j=1}^{Z} \gamma_j \mathcal{I}_j A_{xy}(\mathcal{R}_j), \; M_{yz} = \ldots, \; M_{zx} = \ldots, \tag{18}$$

where $A_{xy}(\mathcal{R}_j), \ldots$, etc. denotes standard area of $\mathcal{R}_j$ and d_z distance of the center of mass from the rotational axis.

This method provides a potentially useful tool for predictive and postdictive diagnostics. By analyzing projected areas, it can be applied to implement tests of accuracy of numerical methods simulating vortex tangles. In superfluids, in particular, by analyzing the area distribution of the vortex projection one can judge about the scale distribution of linear and angular momentum, and compare this with the expected values of the spectrum of turbulence (given, for example, by the Kolmogorov's two-thirds law). Moreover, since LIA preserves an infinity of invariants of motion and *all* of these admit a geometric interpretation in terms of global curvature, torsion and higher order gradients, these can be implemented to supply further information on dynamical properties (for instance, kinetic energy and helicity). Other features associated with the analysis of projected graphs and surface information can be related to dynamical issues: for instance, the Euler characteristic $\chi(G)$ of a graph G associated with a vortex knot or link type. This, being given by $\chi(G) = v - e + r$, where v are the vertices, e the edges and r the regions of G, is a topological invariant related also to the genus $g(F)$ of the Seifert surface F associated with any presentation of the knot or link, by the relation $1 - 2g(F) = \chi(F) = s_0 - c_{\min}$, where s_0 denotes the number of Seifert circles and $c_{\min}$ the topological crossing number of the knot or link. Study of Seifert surfaces of physical systems may reveal interesting properties associated with minimum energy aspects of the system.

4 Topological Bounds on Energy and Helicity-Crossing Number Relations for Magnetic Knots and Links

We restrict our attention to magnetic knots and links: by construction (see, for example, Ricca, 1998) these are tubular embeddings of the magnetic field $\mathbf{B}$ in nested tori T_i ($i = 1, \ldots, n$) centred on smooth, oriented loops χ_i that are knotted and linked in the fluid domain (see Figure 8). We therefore identify

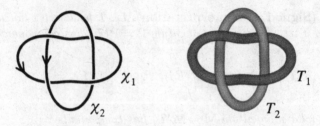

Fig. 8 (a) The 2-component link 4_1^2 with 4 minimum number of crossings is here represented (a) by a disjoint union of 2 oriented loops and (b) by the corresponding centred, tubular neighbourhoods.

an n-component magnetic link $\mathcal{L}_n$ with the standard embedding of a disjoint union of n magnetic solid tori in $I\!\!R^3$:

$$\sqcup_i T_i \;\hookrightarrow\; \mathcal{L}_n := \mathrm{supp}(\mathbf{B}) \;. \tag{19}$$

Let $V = V(\mathcal{L}_n)$ be the total volume of the magnetic link.

We take $\mathbf{B} \cdot \boldsymbol{\nu} = 0$ on each tubular boundary ∂T_i of unit normal $\boldsymbol{\nu}$; the flux $\varPhi_i$ of the magnetic field through each cross-sectional area of T_i is given by:

$$\varPhi_i = \int_{T_i} \mathbf{B} \cdot \boldsymbol{\nu} \, \mathrm{d}^2\mathbf{X} \;. \tag{20}$$

Consider the evolution of $\mathcal{L}_n$ under the action of the group of volume– and flux–preserving diffeomorphisms $\varphi: \mathcal{L}_n \to \mathcal{L}_{n,\varphi}$. Two fundamental physical quantities of the system are the magnetic energy and the magnetic helicity, respectively defined by:

$$M(t) := \int_{V(\mathcal{L}_n)} \|\mathbf{B}\|^2 \, \mathrm{d}^3\mathbf{X} \;, \qquad H(t) := \int_{V(\mathcal{L}_n)} \mathbf{A} \cdot \mathbf{B} \, \mathrm{d}^3\mathbf{X} \;, \tag{21}$$

where $\mathbf{A}$ is the vector potential associated with $\mathbf{B} = \nabla \times \mathbf{A}$. We take $\nabla \cdot \mathbf{A} = 0$ in $I\!\!R^3$.

4.1 Topology Bounds Energy in Ideal Fluid

More specifically, let us consider the class of magnetic fields $\mathbf{B} = \mathbf{B}(\mathbf{X}, t)$ that are solenoidal, frozen and of finite energy in an incompressible and perfectly conducting fluid, that is

$$\mathbf{B} \in \{\nabla \cdot \mathbf{B} = 0, \; \partial_t \mathbf{B} = \nabla \times (\mathbf{u} \times \mathbf{B}), \; L_2\text{-norm}\} \;. \tag{22}$$

For frozen fields helicity is a conserved quantity (Woltjer, 1958), thus $H(t) = H = $ constant. It is known that helicity admits topological interpretation in

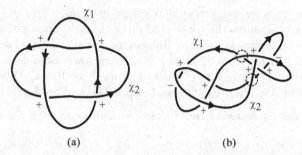

(a)   (b)

Fig. 9 (a) The 2-component oriented link 4_1^2 has 4 minimum number of crossings and Gauss linking number $Lk_{12} = 2$. This, being a topological invariant, does not depend on specific projections: the same link type is shown in (a) its minimal projection, with the 4 crossings denoted by the $+$ sign, and in (b) with redundant crossings. Note that the algebraic sum of signed crossings (omitting self-crossings) remains unchanged: the two crossings in (b), denoted by dashed circles, do not contribute to the linking number calculation of eq. (24) because in each case the crossing strands belong to the same link component.

terms of linking numbers (Moffatt, 1969; Berger & Field, 1984; Moffatt & Ricca, 1992):

Theorem. *Let $\mathcal{L}_n$ be an essential magnetic link in an ideal fluid. Then*

$$H = \sum_i Lk_i \, \Phi_i^2 + 2 \sum_{i \neq j} Lk_{ij} \, \Phi_i \Phi_j \,, \qquad (23)$$

where Lk_i denotes the Călugăreanu-White linking number of χ_i with respect to the framing induced by the embedding of $\mathbf{B}$ in T_i, and Lk_{ij} denotes the Gauss linking number of χ_i with χ_j.

The Gauss linking number Lk_{ij} is a topological invariant of link types and, admitting interpretation in terms of signed crossings (see first of 12 and Figure 9), it can also be written as

$$Lk_{ij} \equiv \int_{i,j} d\omega_{i,j} = \frac{1}{2} \sum_{\substack{r \in \chi_i \sqcap \chi_j \\ i \neq j}} \epsilon_r \,, \qquad (24)$$

where $d\omega_{i,j}$ is the classical Gauss integrand form associated with the two curves χ_i, χ_j and $\sqcap$ denotes apparent intersections of curve strands, omitting self-crossings. The Călugăreanu-White linking number Lk_i is a topological invariant of each link component and admits a geometric decomposition in terms of writhing number Wr_i and twist number Tw_i, according to the well-known formula (Călugăreanu, 1961; White, 1969):

$$Lk_i = Wr_i + Tw_i \,. \qquad (25)$$

The writhing number measures the average distortion of χ_i in space, while the total twist measures the total winding of the field lines within T_i.

Assuming for simplicity that all link components have equal flux Φ and be zero-framed, that is $Lk_i = 0$ for all $i = 1, \ldots, n$, lower bounds on energy are given by the following results (for detailed proof see Ricca, 2008a, based on previous works done by Arnold, Freedman & He, and Moffatt):

Theorem. *Let $\mathcal{L}_n$ be an essential magnetic link in an ideal fluid. Then*

$$(i) \quad M(t) \geq \left(\frac{16}{\pi}\right)^{1/3} \frac{|H|}{V^{1/3}} \; ; \qquad (ii) \quad M_{\min} = \left(\frac{16}{\pi}\right)^{1/3} \frac{\Phi^2 c_{\min}}{V^{1/3}} \; , \qquad (26)$$

where $c_{\min}$ is the topological crossing number of $\mathcal{L}_n$.

The theorem above establishes two important results: (26-i) states that magnetic energy is bounded from below by the absolute value of the helicity (given by the total linking number of the system), scaled by the average size of the system: hence, high helicity concentrations imply greater bounds on energy values. Equality (26-ii) states that the energy minima are actually given by the topological crossing number of the system scaled by the system average size. Note, however, that a classification of knots and links based on energy contents is still incomplete, without prescribing individual framing. By direct inspection of link tabulation, it is indeed immediately evident that there are countably many topologically distinct links with equal number n of (zero-framed) components and same $c_{\min}$ (see Figure 10). This means that a complete classification of magnetic systems by topology is only possible by specifying the individual framing of each component.

4.2 Helicity-Crossing Number Relations in Dissipative Fluid

Suppose now that the fluid is no longer perfectly conducting, but resistive, assuming that the typical dissipation (or reconnection) time scale is higher than the typical evolution time scale so as to preserve magnetic flux. The topology of $\mathcal{L}_n$ may now change due to the effects of dissipation, that make reconnections of the magnetic field lines possible. Under these conditions magnetic helicity may also change, hence $H = H(t)$. Let us define $\Omega := \sum_{\{i,j\}} d\omega_{i,j}$; the change in magnetic helicity can be measured in terms of change in algebraic complexity of the magnetic link according to the following result (for proof see Ricca, 2008a).

Theorem. *Let $\mathcal{L}_n$ be a zero-framed, essential magnetic link, embedded in a resistive, incompressible fluid. Then, we have:*

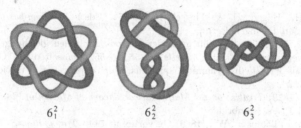

$$6_1^2 \qquad\qquad 6_2^2 \qquad\qquad 6_3^2$$

Fig. 10 Three distinct link types with $n = 2$ and $c_{\min} = 6$. By assuming zero-framing in all link components, same volume V and flux Φ, the three links must have same groundstate energy $M_{\min}$. Thus, different framing should be prescribed if we want to identify uniquely each knot/link type with its specific groundstate energy.

$$\text{(i)} \quad |H(t)| \le 2\Phi^2 \overline{C}(t) \; ; \quad \text{(ii)} \quad \frac{d|H(t)|}{dt} \le \text{sign}(H)\text{sign}(\Omega)2\Phi^2 \frac{d\overline{C}(t)}{dt} \; . \quad (27)$$

Direct numerical tests on tangle complexity (cf. results shown in Figure 6; see Barenghi *et al.*, 2001) confirm these results: from the data analysis of the tangle mature growth stage (Φ in appropriate units), we have $[2\overline{C}(0.09) - |H(0.09)|]/|H(0.09)| \approx 19.3\%$ and $\text{sign}(H)\text{sign}(\Omega) = +$, $[2\Delta\overline{C}(t) - \Delta|H(t)|]/\Delta|H(t)| \approx 27.6\%$ for $t \in [0.08, 0.09]$. Since these results are independent of specific viscous or resistive time scales, physical time ought to be interpreted in terms of the reconnection time scale involved in the change of topology.

Acknowledgements Financial support from Italy's MIUR (D.M. 26.01.01, n. 13 "Incentivazione alla mobilità di studiosi stranieri e italiani residenti all'estero") and from ISI-Fondazione CRT (Lagrange Project) is kindly acknowledged.

References

1. Barenghi C.F., Ricca, R.L. & Samuels D.C. (2001) How tangled is a tangle? *Physica D* **157**, 197–206.
2. Barenghi C.F., Ricca, R.L. & Samuels D.C. (2002) Complexity measures of tangled vortex filaments. In *Tubes, Sheets and Singularities in Fluid Dynamics* (ed. K. Bajer & H.K. Moffatt), pp. 69-74. NATO ASI Series, Kluwer.
3. Berger, M.A. & Field, G.B. (1984) The topological properties of magnetic helicity. *J. Fluid Mech.* **147**, 133–148.
4. Bott, R. & Tu, L.W. (1982) *Differential Forms in Algebraic Topology.* Graduate texts in Mathematics **82**, Springer, Berlin.
5. Călugăreanu, G. (1961) Sur les classes d'isotopie des nœuds tridimensionnels et leurs invariants. *Czechoslovak Math. J.* **11**, 588–625.
6. Chong, M.S., Perry, A.E. & Cantwell, B.J. (1990) A general classification of three-dimensional flow fields. *Phys. Fluids* A **2**, 765–777.
7. Freedman, M.H. & He, Z.-X. (1991) Divergence-free fields: energy and asymptotic crossing number. *Ann. Math.* **134**, 189–229.

8. Hauser, H., Hagen, H. & Theisel, H. (2007) (Eds.) *Topology-based Methods in Visualization.* Springer, Berlin.

9. Helmholtz, H. (1858) Über integrale der hydrodynamischen gleichungen welche den wirbelbewegungen entsprechen. *Crelle's J.* **55**, 25–55. [Transalted by P.G. Tait: (1867) On integrals of the hydrodynamical equations, which express vortex motion. *Phil. Mag.* **33**, 485–512.]

10. Lamb, H. (1879) *Treatise on the Mathematical Theory of Motion of Fluids.* Cambridge University Press, Cambridge.

11. Lord Kelvin (Thomson, W.) (1869) On vortex motion. *Trans. Roy. S. Edinburgh* **25**, 217–260.

12. Ma, T. & Wang, S. (2005) *Geometric Theory of Incompressible Flows with Applications to Fluid Dynamics.* Mathematical Surveys and Monographs **119**, American Mathematical Society.

13. Maxwell, J.C. (1873) *A Treatise on Electricity and Magnetism.* Clarendon Press, London.

14. Moffatt, H.K. (1969) The degree of knottedness of tangled vortex lines. *J. Fluid Mech.* **35**, 117–129.

15. Moffatt, H.K. & Ricca, R.L. (1992) Helicity and the Călugăreanu invariant. *Proc. R. Soc.* A **439**, 411–429.

16. Ricca, R.L. (1998) Applications of knot theory in fluid mechanics. In *Knot Theory* (ed. V.F.R. Jones *et al.*), pp. 321–346. Banach Center Publs. **42**, Polish Academy of Sciences, Warsaw.

17. Ricca, R.L. (2000) Towards a complexity measure theory for vortex tangles. In *Knots in Hellas '98* (ed. C. McA. Gordon *et al.*), pp. 361–379. Series on Knots & Everything **24**, World Scientific, Singapore.

18. Ricca, R.L. (2001) Tropicity and complexity measures for vortex tangles. In *Quantized Vortex Dynamics and Superfluid Turbulence* (ed. C.F. Barenghi *et al.*), pp. 366–372. Springer Lecture Notes in Physics **571**, Springer, Berlin.

19. Ricca, R.L. (2005) Structural complexity. In *Encyclopedia of Nonlinear Science* (ed. A. Scott), pp. 885–887. Routledge, New York and London.

20. Ricca, R.L. (2008a) Topology bounds energy of knots and links. *Proc. R. Soc.* A **464**, 293–300.

21. Ricca, R.L. (2008b) Momenta of a vortex tangle by structural complexity analysis. *Physica* D, in press. (doi:10.1016/j.physd.2008.01.002).

22. Riemann, B. (1857) Lehrsätze aus der analysis situs für die theorie der integrale von zweiigliedrigren vollständingen differentialien. *Crelle's J.* **54**, 105–110.

23. Saffman, P.G. (1991) *Vortex Dynamics.* Cambridge University Press, Cambridge.

24. Weickert, J. & Hagen, H. (Eds.) (2006) *Visualization and Processing of Tensor Fields.* Springer, Berlin.

25. Weintraub, S.H. (1997) *Differential Forms.* Academic Press Inc., San Diego.

26. White, J.H. (1969) Self-linking and the Gauss integral in higher dimensions. *Amer. J. Math.* **91**, 693–728.

27. Woltjer, L. (1958) A theorem on force-free magnetic fields. *Proc. Natl. Acad. Sci. USA* **44**, 489–491.

Random Knotting: Theorems, Simulations and Applications

De Witt Sumners
(CIME Lecturer)

Abstract This article describes some of the theoretical and simulation results on random entanglement, and give a few scientific applications. I will prove that, on the simple cubic lattice Z^3, the probability that a randomly chosen n-edge polygon in Z^3 is knotted goes to one exponentially rapidly with length n (Murphy's Law of entanglement); in other words, all but exponentially few polygons of length n in Z^3 are knotted. Measures of entanglement complexity of random knots and random arcs are discussed as well as application of random knotting to viral DNA packing.

1 Introduction

A fundamental problem in science is the description and quantification of topological entanglement complexity of filaments — embedded arcs, circles and graphs in 3-space. This problem is the province of the mathematical subject of *knot theory*, and it has a fascinating history going back to the work of Gauss and Maxwell [Sv]. In more recent times, physicists and mathematicians (stimulated by the behavior of real filament systems), have become interested in measuring entanglement of random systems and entanglement changes with length and/or density (length per unit volume), and in understanding the physical ramifications of entanglement. Random entanglement increases with filament length and/or density; physical intuition is absolutely clear on this point. As (almost) everyone has experienced, a 100 foot electrical extension cord (when carelessly bunched up and put in a corner of the garage by your teenager) is difficult to untangle when you get ready to

De W. Sumners
Department of Mathematics, Florida State University, Tallahassee,
FL 32303-4510, USA
e-mail: sumners@math.fsu.edu

R.L. Ricca (ed.), *Lectures on Topological Fluid Mechanics*,
Lecture Notes in Mathematics 1973, DOI: 10.1007/978-3-642-00837-5,
© Springer-Verlag Berlin Heidelberg 2009

use it — when you find the ends of the cord and pull them apart, the cord is invariably highly entangled, and has at least one knot in it. In order to use the extension cord, you must snake a cord end through the tangle to resolve it. By comparison, a 25 foot extension cord (stored in the same careless fashion) untangles much more easily — you need only to find the ends and pull them apart, often outstretching the cord with little or no entanglement. Can one quantify observations like this, and/or prove theorems about entanglement? The answer is YES, and this article will describe some of the theoretical and simulation results on random entanglement, and give a few scientific applications. This article is by no means exhaustive; I will only describe some of the work I am most familiar with. I will, for example, prove that, on the simple cubic lattice Z^3, the probability that a randomly chosen n-edge polygon in Z^3 is knotted goes to one exponentially rapidly with length n; in other words, all but exponentially few polygons of length n in Z^3 are knotted. I once had the experience of presenting this proof (Murphy's Law of entanglement) at a research conference; one of the questions I was asked after my talk (by a non-mathematician!) was "Why does one need to prove that the longer a random polygon, the more likely it is to be knotted? This phenomenon is obvious, and needs no proof!" Applying that same line of reasoning, the Jordan-Schoenflies Curve Theorem (which says that every circle in the XY-plane bounds a compact region (the inside) homeomorphic to the 2-disk) is also obvious and needs no proof. As every mathematician knows, not every "obvious" statement is true, and a proof is the intellectual laboratory where mathematical truth can be verified. An interesting example of something which is "obvious" but in fact false is the 3-dimensional analogue of the Jordan-Schoenflies Theorem — that every 2-sphere in 3-space bounds a 3-disk. This was thought to be true by mathematicians, and Alexander even claimed a proof, but took it back when he found a mistake in his proof, and later a counterexample — the infamous Alexander Horned Sphere [Ax1]. The horned sphere is very badly embedded in 3-space — it has a Cantor set of points where the embedding is "wild". By restricting to finite polyhedral 2-spheres (which cannot have any "wild" points), Alexander was able to prove the 3-dimensional version of the Jordan-Schonflies Theorem [Ax2].

What are the effects of entanglement in real physical systems? Leaving aside your frustration at resolving the entanglement in the spaghetti-like mass of computer wires under your desk, unresolved entanglement of DNA in cells is a death sentence for that cell [BZ, LZC]. Most drugs for the treatment of bacterial infections or cancer work by inhibiting cellular enzymes, which resolve molecular entanglement in the cell, and the target cell (a pathogen or cancer cell) dies as a result. In polymer science, macroscopic properties of polymer systems often depend on microscopic intermolecular entanglement [RBHS]; entanglement determines whether or not the polymer system is a gel or a polymer fluid, and if a solid, the entanglement has consequences in the strength of the material. In fluid dynamics, plasma and superfliud physics [Mof,Ric1,Ric2], entanglement of magnetic and vortex filaments have important consequences for the energy of the system.

2 The Frisch-Wasserman-Delbruck Conjecture

The specific problem of occurrence of knots in a long linear polymer chain was first addressed independently by Frisch and Wasserman in 1961 [FW,Was], and by Delbruck in 1962 [Del]. Both groups formulated questions about the probability that a closed circular polymer chain with degree of polymerization n (n monomers) would contain a knot. More specifically, if one performs a cyclization reaction (random closure) on a dilute solution containing linear polymers with polymerization degree n, what would be the yield of this reaction? Neglecting dimers, trimers, etc., and focusing only on the circular reaction products with n monomers, what is the product spectrum (histogram of knot types)? Frisch and Wasserman and Delbruck conjectured (no surprise here!):

Conjecture 2.1 (Frisch-Wasserman Delbruck (FWD) Conjecture). The probability that a randomly embedded circle of length n in R^3 is knotted tends to one as n tends to infinity.

Numerical evidence pointing to the truth of the FWD conjecture abounds: The first Monte Carlo simulation of knotting in random polygons was done by Vologodskii *et al.* [VLF]. They generated a random sample of polygons of length n, and used $\Delta(-1)$, the *order of the knot* (the Alexander polynomial evaluated at $t = -1$) to detect knotting. One of the models they investigated was a random walk model on the simple cubic lattice Z^3. One starts at the origin, and performs a random walk; when a self-intersection of the walk is encountered, perturb the entire lattice a small amount in a random direction to remove the intersection and keep walking. As one continues on the walk, bias the walk to return to the origin after n steps. They found that the knot probability rose with increasing length n, and obtained quite high knot probabilities (60% for $n = 300$). Other groups [CM,MW] performed similar studies 20-30 years ago, with similar results — the knot probability grows with n, tending toward unity as n tends to infinity.

In 1986 I gave a talk at a Canadian Chemical Society meeting in Saskatoon where I presented the FWD conjecture. Stu Whittingon was in the audience, and after my talk came up and asked "Have you heard of the Kesten Pattern Theorem?" At that point Stu and I began to work on the FWD conjecture. The model we chose to use was self-avoiding walks (SAW) and self-avoiding polygons (SAP) on Z^3. The simple cubic lattice is useful for describing excluded volume effects in polymers in dilute solution, and allows one to do both rigorous asymptotic proofs as n goes to infinity, and numerical simulations for small values of n, allowing us to complete a proof of the FWD conjecture [SW].

On the simple cubic lattice Z^3, a step is a directed edge joining two adjacent lattice points. An *n-step self-avoiding walk* (n-SAW) beginning at lattice point x_0 is an $(n+1)$-tuple of distinct lattice points $\{x_1, \ldots, x_n\}$ where x_i and x_{i+1} are adjacent in the lattice for $0 \leq i < n$. An n-step self-avoiding polygon (n-SAP) is an n-SAW whose first and last vertices are adjacent in the lattice. An n-SAW (n-SAP) is *rooted* if $x_0 = 0$. Since there are finitely many rooted n-SAWs (n-SAPs) for each n, then the probability that a randomly chosen n-SAP is knotted is simply the ratio of the number of rooted, knotted n-SAP divided by the number of rooted n-SAP.

The key idea in the proof of the FWD conjecture is the Kesten Pattern Theorem [Kes]. A Kesten pattern X is a SAW which has a way in and a way out, a walk that can be concatenated so that it can appear many times in a long SAW. For an example of a SAW in Z^3 which is not a Kesten pattern, consider a "crab trap", a lattice cube in which the boundary sphere of the cube is saturated by the walk, which then enters the interior of the cube. Once inside the cube, the walk cannot exit the cube because all the boundary vertices are already occupied, and the walk must terminate inside the cube. Kesten proved that given any Kesten Pattern X there is a positive density $D_X > 0$ associated to this pattern such that, for sufficiently large n, X appears (up to translation) at least $\lfloor D_X n \rfloor$ times in all but exponentially few self-avoiding walks of length n. The first step was to extended Kesten's result to cover patterns in SAP as well as SAW on Z^3 [SW].

So a Kesten pattern appears at least once in all but exponentially few sufficiently long SAW and SAP. In order to prove the FWD on Z^3, one needs to produce a Kesten pattern T such that if T appears in a given SAP, then that SAP is guaranteed to be knotted. A *tight knot* is such a pattern, for example the one specified by the SAW T given below. Suppose that we have a right-handed coordinate system in Z^3, and let i, j, k be the unit vectors in the X, Y, Z directions, respectively. Beginning at the origin in Z^3, take the following 18-step walk:

$$T : \{j, j, -i, k, k, i, i, -k, -j, -k, -k, -i, -i, k, k, i, j, j\}.$$

How can we be guaranteed that if the SAW T appears in a SAP, then that SAP is knotted? Ordinarily local patterns such as T in a long circle to not guarantee that the circle is knotted, because the circle can snake back through the local entanglement T and undo the local knot (just as one resolves the entanglement in a long extension cord). However, the self-avoiding condition prevents any such snaking back through the entanglement, and any SAP that contains the pattern T must be knotted. More precisely, each occupied vertex in a SAW (SAP) sits in the middle (barycenter) of a dual 3-cube, and one can think of this dual 3-cube as the excluded volume generated by that occupied lattice site. Let T be the 16-step sub walk of T obtained by deleting the first and last steps. Let $N(T')$ denote the *lattice neighborhood* of T' — the union of the 16 dual 3-cubes which surround the vertices of T'. $N(T')$ is a 3-ball, and T enters and exits $N(T')$ transversely in its first and last steps. Suppose now that K is any SAP that contains T. The 2-sphere boundary of $N(T')$ separates K into the connected sum of two knots, one of which is the trefoil formed by the intersection of K and $N(T')$. In order to prove that K is knotted, we compute the *genus* of K. Every knot K is spanned by many orientable surfaces (called *Seifert surfaces*). If one takes the minimum genus over all the Seifert surfaces spanning the knot, one obtains the *genus* of the knot $g(K)$. K is unknotted if and only if K spans a 2-disk, so K is unknotted if and only if $g(K) = 0$, and K is knotted if and only if $g(K) \geq 1$. Since the knot genus is additive on connected sums [Adm], and the trefoil is a summand of K of genus one, then the genus of K is at least one, hence K is knotted. A similar proof of the FWD conjecture (also based on Kesten's pattern theorem) was found independently by Nick Pippenger [Pip].

The number p_n of rooted SAP of length n in Z^3 behaves as [RSW]

$$p_n = e^{\kappa n + o(n)}$$

and the number of unknotted polygons p_n^0 behaves as

$$p_n^0 = e^{\lambda n + o(n)}$$

with $0 < \lambda < \kappa$ so that the knot probability $P(n)$ behaves as

$$P(n) = 1 - e^{-\alpha n + o(n)}$$

for some positive constant α.

Theorem 2.2 (SW, Pip). *The probability that an n-SAP in Z^3 is knotted goes to one exponentially rapidly as n tends to infinity.*

Although we have proved that there exists a positive constant that describes the knotting probability $P(n)$, no rigorous analytic method for computing is known, and the value of must be determined by simulation. For $n < 24$, there are no knotted SAPs, and for $n = 24$, there are exactly 3496 knotted 24-SAPs (all trefoils) [Do1], out of a total of something of the order of 10^{13} 24-SAPs, so $P(n) = 0$ for $n < 24$ and $P(24)$ is positive but vanishingly small. Simulation results on Z^3 produce something on the order of 1% knots for $n = 1000$ [RW]. Knot-type specific estimates of knot probability parameters have also been made [DT1].

We have now proved that all but exponentially few sufficiently long SAP contain a tight trefoil. What about other knots? In fact, the above argument can be used to show that, as n tends to infinity, every knot eventually appears as a summand — that is, if K^* is a fixed knot type, then all but exponentially few sufficiently long SAP contain a copy of K^* as a summand [SSW]. Let K^* denote a knot type in S^3, and let K^* be a polygonal representative of K^*

in S^3. Insert a new vertex in the interior of an edge of K^*, and regard this new vertex as the point at infinity in S^3. By removing this point, we get a knotted polygonal arc in R^3 with the property that the ends of the knotted arc go off to infinity along the X-axis. Take a regular projection of this arc on the XY ($z = 0$) plane; by forgetting the crossover information (over-under at each crossing), and using isotopy in the XY plane and subdivision when necessary, we obtain an immersion of the arc in the square lattice XY plane, and the ends go off to infinity along the X axis. The non-trivial part of the immersed arc (all of the immersion except for parts of the ends which go off to infinity) is contained in a square in the XY plane with vertices $(\pm d, \pm d, 0)$ for some even integer d. This square intersects the immersion in two points where the straight ends of the immersed arc go off to infinity. We can now recover the knotted arc by remembering the crossing information. Replace each two lattice steps which represent an underpass by a path which detours one step into the $z = -1$ plane, continues two steps as the underpass in the $z = -1$ plane, and then returns in one step to the $z = 0$ plane. We now have an infinite SAW contained in two parallel planes, $z = 0$ and $z = -1$. This arc exits the square with vertices $(\pm d, \pm d)$ at two points: $(-d, 0, 0)$ and $(d, 0, 0)$. By removing the infinite ends (remove all vertices $(x, 0, 0)$ with $x \leq -d$ and $x \geq d$), one obtains the finite SAW α where α has captured the knot type K^*. The lattice neighborhood $N(\alpha)$ is homeomorphic to a punctured 2-disk cross an interval — a piece of Swiss cheese. We need to fill in the holes by extending α to α^* such that α^* still represents the knot K^* and $N(\alpha^*)$ is a 3-ball. We do this by draping the ends of α over the entire disk, analogous to pouring syrup on a pancake, filling up all the holes. Start at the vertex $(d, 0, 0)$, the right-hand endpoint of α. Add the vertex $(d, 0, 1)$, then traverse upward through $(d, 2, 1), (d, 3, 1), \ldots, (d, d, 1)$. Proceed left to vertex $(d-1, d, 1)$, then traverse down through $(d - 1, d - 1, 1), \ldots, (d - 1, -d, 1)$; then left to vertex $(d-2, -d, 1)$, then up through $(d-2, -d+1, 1)$, etc. Proceed to zigzag up and down until vertex $(1, d, 1)$ is reached (on an upward traversal), then add an endpoint vertex $(1, d + 1, 1)$. These steps which have been added to α drape the SAW over the right-hand half of the enclosing square, and the knot type has been preserved because the added steps can be vertically pulled up and off of the "knotted" part of the arc. Perform the analogous procedure on the left-hand half of the square, starting with the vertex $(-d, 0, 0)$, moving one step up to the $z = 1$ plane, then proceeding up the left-hand edge of the square until the top is reached, then moving one step to the right, going straight down until the bottom of the square is reached, then up, etc. On the last downward traversal, the vertex $(0, -d, 1)$ is reached, and one then adds an endpoint $(0, d, -1)$. The SAW α^* so produced is the desired Kesten pattern representing the knot type K^*.

We have shown the following:

Theorem 2.3 (SSW). *Let K denote a given knot type; then there exists a positive density D_K such that, for sufficiently large n, the knot K appears at* *least* $\lfloor nD_K \rfloor$ *times as a summand of all but exponentially few SAPs of length* n *on* Z^3.

There are two other models in which the FWD conjecture has been proven. One model is the Gaussian Random Polygon (GRP) model. An n-GRP is a piecewise linear circle in R^3 with n edges in which the edge lengths form a Gaussian distribution. In [DPS] a continuum version of the Kesten pattern theorem was proved, and then used to prove that the probability that a randomly chosen n-GRP is knotted tends to one exponentially rapidly as n tends to infinity. The other is the Equilateral Polygon (EP) model. An n-EP is an equilateral polygon in R^3 with n edges. In [Do2] Diao proved the FWD conjecture for equilateral polygons. All of the proofs of the FWD discussed to this point have relied on local (tight) knots to force knotting of sufficiently long polygons. Can it be shown that almost all sufficiently long random polygons have global (non-local) knots in them? The answer is YES. Jungreis [Jun] proved global knotting in the GRP model; with high probability a long randomly chosen polygon is a satellite knot (it is an essential loop in a knotted solid torus), so the knot cannot be unknotted by small perturbations which could kill local knots). Diao *et al.* [DNS] also proved global knotting in the EP model. As far as I know, global knotting for SAP on Z^3 is almost certainly true but has not yet been proved.

The codimension two phenomenon of knotting occurs in all dimensions; p-spheres can be knotted in $(p+2)$ space [Rol] for all $p \geq 1$. Consider rooted p-spheres in Z^{p+2}. Consider p-spheres in Z^{p+2} that are the union of *unit* p-cells, where a *unit* p-cell in Z^{p+2} is one spanned by 2^p vertices that are adjacent in Z^{p+2}. Let $n - S^p$ denote a p-sphere with n unit p-cells. The integer n is the n-dimensional area measure of the p-sphere. There are finitely many rooted $n - S^p$ for each n. The generalization of the FWD to higher-dimensional knots is:

Conjecture 2.4 (Generalized FWD Conjecture). The probability that a randomly chosen $n - S^p$ in Z^{p+2} $(p \geq 2)$ is knotted tends to one as n tends to infinity.

In the above conjecture, the number n represents the p-dimensional volume measure (the size) of the p-sphere in Euclidean $(p + 2)$-space. The above conjecture of the inevitability of knotting of randomly embedded spheres as the volume of the sphere tends to infinity can be made in any codimension 2 context in Euclidean space; smooth codimension 2 spheres, piecewise linear codimension 2 spheres, locally flat codimension 2 spheres, etc. Efforts to prove this conjecture for $p = 2$ (2-spheres in 4-space) have been made, without success. One problem is that there is no known analogue of a Kesten Pattern Theorem for 2-disks in Z^4, so a new idea may be required in order to prove this conjecture. To my knowledge, although almost certainly true, no numerical simulation evidence for this conjecture exists.

3 Entanglement Complexity of Random Knots and Random Arcs

Intuition tells us that the entanglement of random knots grows with length — the longer it is, the more entangled. We are now positioned to measure entanglement complexity, and prove that the complexity grows at least linearly with the length. The reason for this growth in complexity with length is that long random knots are highly composite (have many summands), and most measures of knot complexity are additive under connected sum. It is instructive at this point to recall the work of Kendall [Ken] on the complexity of Brownian motion. Kendall proved that given any (smooth or finite polygonal) arc in R^3, and given any closed tube neighborhood of this arc, the Brownian motion eventually enters one end of the tube for the last time, traverses around the interior of the tube, then exits the other end of the tube. This happens no matter how complicated the arc and how small the tube diameter. So, Kendall has shown that Brownian motion exhibits *all knots at all scales*.

Let $\mathcal{K}$ denote the set of knot types in R^3. A *good measure of knot complexity* is a function $F : \mathcal{K} \to [0, \infty)$ that satisfies the following:

(a) $F(\text{unknot}) = 0$
(b) There exists a knot type $K \in \mathcal{K}$ such that $F(nK\sharp L) \geq nF(K) > 0$ for all $L \in \mathcal{K}$, where $\sharp$ denotes connected sum of knots.

Good measures of knot complexity are designed so that the complexity of any knot that contains pK as a summand is bounded below by p times the complexity of K; hence any good measure of knot complexity diverges to infinity at least linearly with length. More precisely, we have the following lemma:

Lemma 3.1 (SSW). *For any F (a good measure of knot complexity), let K be a knot that satisfies part (b) of the definition above. Then, there exists a positive integer n_K such that for sufficiently large $n > n_K$, all but exponentially few n-SAPs have F-complexity which exceeds $F(K)\left((n/n_K) - 1\right)$.*

Proof. Choose D_K as in Theorem 2 above and choose n_K such that $\lfloor (n_K - 1)D_K \rfloor = 0$ and $\lfloor n_K D_K \rfloor = 1$. For sufficiently large $n > n_K$, all but exponentially few n-SAPs K' are of the form $K' = \lfloor n_K D_K \rfloor K \sharp L$ for some $L \in \mathcal{K}$. This means that $F(K') \geq \lfloor n_K D_K \rfloor F(K) > F(K)((n/n_K) - 1)$. $\square$

Theorem 3.2 (SSW). *The following are good measures of knot complexity:*

(a) number of prime factors
(b) genus
(c) bridge number -1
(d) span of any non-trivial knot polynomial
(e) log(order)

(f) crossing number
(g) unknotting number
(h) minor index
(i) braid index −1

Proof. All of the above are non-trivial non-negative integer knot invariants, and some are known to be additive on connected sum. In any event, each of them is additive on trefoil summands of a given knot, so the fact that random knots tend to have many trefoil summands means that their complexity must grow at least linearly with length. I will give the argument for one of the more interesting entanglement invariants, the unknotting number. For a knot K, the *unknotting number* $\mu(K)$ is the minimum number of times that the knot must be passed through itself in order to unknot it. Unknotting number is believed to be additive on connected sums:

Conjecture 3.3 (Additivity of Unknotting Numbers). $\mu(K_1 \sharp K_2) = \mu(K1) + \mu(K_2)$.

For any knot K, let X denote the bounded knot complement in S^3, and X^* denote the infinite cyclic covering space of X. $H_1(X^*; Z)$ is presented as a module over the ring $\Lambda = Z[t, t^{-1}]$ (the integral group ring of the infinite cyclic multiplicative group) by a square matrix with entries in Λ, called the *Alexander matrix*. Let $m(K)$ denote the *minor index* of K, that is, the minimum size for any square Alexander matrix that presents the Λ-module $H_1(X^*; Z)$.   □

We now use the following theorem of Nakanishi:

Theorem 3.4 (Nak). *For any* $K \in \mathcal{K}, 0 \le m(K) \le \mu(K)$.

Hence, if $K = 3_1$ (the trefoil knot), then $0 \le n \le m(nK \sharp L) \le \mu(nK \sharp L)$ for any $L \in \mathcal{K}$, so the minor index and the unknotting number are good measures of knot complexity.   □

Scientists continue to be interested in measuring entanglement of arcs in 3-space — they would like to be able to measure exactly where an arc is knotted, for example. In all of 3-space, every arc is unknotted — one can thread an end through any entanglement in order to resolve it. However, if the ends are constrained (to lie in a 2-sphere for example), the arc can be knotted. In the proof of the FWD conjecture above, the pair $(N(T')$, $N(T') \cap T)$ is a *knotted ball pair* — a knotted arc in an enclosing 3-ball. On the other hand, if one were to take a closed tube neighborhood B of the SAW T (with the ends of T in the ends of B), then the pair (B, T) would be an unknotted ball pair — whether or not the arc T is unknotted depends on the surrounding 3-ball one chooses [SW, Fig. 1].

Given any (smooth, finite polygonal) arc (circle) in 3-space, it is possible to measure the (geometric) complexity of the embedding as follows: given any regular projection of the arc (circle), compute the crossing number, and then average this quantity over all projections, to get the *average crossing number* of the arc (circle). It is possible to alter the trefoil pattern T by adding edges to obtain a slightly longer trefoil pattern S such that one sees at least 3 crossings in every regular projection of S. The Kesten pattern theorem guarantees that all but exponentially few sufficiently long n-SAWs (n-SAPs) contains at least $\lfloor nD_S \rfloor$ copies of the trefoil pattern S, so the average crossing number of any such n-SAW (n-SAP) is bounded below by $3\lfloor nD_S \rfloor$. This means that if $\mathcal{X}_n$ denotes the expected value of average crossing number over all n-SAWs (n-SAPs), we have the following result:

Theorem 3.5 (RSW). $\mathcal{X}_n \to \infty$ *with* n, *and the divergence is at least linear.*

In order to detect true topological entanglement in an arc in 3-space, one needs to join the ends up to form a circle, and then use topological measures of entanglement for circles. In [RSW] this closing operation was performed on n-SAWs in Z^3 using two methods: (a) Choose a direction at random, and construct two parallel rays from the endpoints of the SAW to infinity. One can assume that the rays meet at the point infinity in S^3, so this operation produces a circle. (b) Choose a direction at random, and add two short (length less than one) parallel line segments to the ends of the n-SAW, then close up by adding a straight line segment connecting the ends of the added segments. This almost always produces an embedded circle. It can be shown [RSW] that either of these closure methods produces circles whose (topological) entanglement complexity (as in Theorem 4) diverges to infinity at least linearly with n. In [RSW] the results of Monte Carlo simulations are discussed which show that for $n \leq 2000$, the entanglement measure expected log(order) gives almost identical results for both methods of n-SAW closure, and that this measure grows with increasing n.

4 Writhe, Signature and Chirality of Random Knots

There are other interesting measures of geometric and topological entanglement complexity which are additive when applied to connected sums of knots, but can take on negative as well as positive values, so it is possible for the measure to yield zero when evaluated on a nontrivial connected sum. For measures such as this, we will show that a lower bound for the growth of the absolute value of one of these measures is $\sqrt{n}$ (instead of n itself).

Given a regular planar projection in the direction ξ of the (smooth, finite polygonal) oriented circle K in R^3, assign the integer ± 1 to each crossing as determined by the right-hand rule (*oriented skew lines sign convention*) in Figure 1. Note that the sign of a given crossing in the projection is independent of orientation of the circle; reversing the orientation of the circle in turn reverses the orientation of each arrow in a crossing, leaving the sign

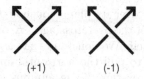

Fig. 1 Oriented skew lines sign convention.

of the crossing invariant. Adding up the signs for each crossing in the regular projection, one obtains the *projected writhe* $\omega_\xi(K)$; by averaging over all directions $\xi \in S^2$, we obtain the *writhe* $\omega(K)$. The writhe is a real-valued geometric measure of non-planarity of K; if K is a subset of a plane in R^3, then $\omega(K) = 0$ [Ful]. Note that $\omega(K)$ is not a topological invariant — the writhe changes when one twists or bends a knot.

In order to approximate the writhe for a given knot K, one needs to choose a finite (but usually large) set of directions, compute the projected writhe for each direction, then average the results. For SAPs on Z^3, however, one can compute the writhe exactly using only 4 directions because of the symmetry of Z^3 [LS,Cim,LaS]. Moreover, the (first) proof of this fact [LS] used linking numbers to characterize the writhe; the topological invariance of linking numbers and the fact that one need use only 4 directions is useful both in proving rigorous results about the writhe of random n-SAPs and in doing Monte Carlo simulations to compute average properties of the writhe for n-SAPs [ROS].

Suppose that we are using the projection direction ξ (ξ is a unit vector); let K_ξ denote the pushoff of K in the direction ξ; $K_\xi = K + s\xi$, where s is a small positive number. The pushoff K_ξ inherits its orientation from K. Let $Lk(K, K_\xi)$ denote the *linking number* of K with K_ξ.

Lemma 4.1 (LS). $\omega_\xi(K) = Lk(K, K_\xi)$.

Proof. The linking number $Lk(K, K_\xi)$ is computed from any regular projection of the pair of oriented curves (K, K_ξ). Given any regular planar projection, ignore self-crossings of each curve, and assign the integer ± 1 to each crossing involving the top strand from K and the bottom strand from K_ξ. The sum of these signed crossings is the linking number. We would like to use the projection in direction of unit vector ξ to determine the linking number; of course, since K_ξ is the pushoff of K in direction ξ, K_ξ lies in the shadow of K and we cannot see it at all in the projection. However, in the plane of the projection, as one walks around the knot diagram in the orientation direction, one can construct the *planar pushoff* of the projection of K; push K to the right to obtain a parallel copy K^* of K. K and K^* form a pair of railroad tracks; each original crossing of K becomes 4 crossings where the railroad crosses itself. By putting a small upward bump in the overcrossing track, we see that there is a small embedded annulus (horizontal curtain) connecting K to K^*. Moreover, each original overcrossing of K gives rise to a single overcrossing of K over its planar pushoff K^*. Since K^* is oriented in parallel to K, the sign of the overcrossing of K over K^* is identical to the sign of K overcrossing itself. We conclude that $\omega_\xi(K) = Lk(K, K^*)$. Now allow the force of gravity to pull the horizontal annulus curtain to vertical position. This downward rotation of the annulus is an isotopy in the complement of K which takes the planar pushoff K^* to the pushoff K_ξ. Hence, $Lk(K, K^*) = Lk(K, K_\xi)$, and we conclude that $\omega_\xi(K) = Lk(K, K_\xi)$. $\square$

Consider now the 2-sphere S of projection (pushoff) directions in R^3, and take S to have radius $1/2$, centered at the origin. The 3 coordinate planes in R^3 separate S into 8 connected regions (octants) specified by constancy of sign in each coordinate. The interior of an octant consists of points with no coordinate $= 0$. Let K be a rooted SAP in Z^3. For the vector direction $\xi \in S$, we let $K_\xi = K + \xi$ denote the pushoff of K in direction ξ.

Claim (1). If ξ lies in the interior of any octant, then K and K_ξ are disjoint. Without loss of generality, assume that ξ lies in the interior of the octant where all 3 coordinates are positive. Then $\xi = (\xi_1, \xi_2, \xi_3)$ where $0 < \xi_i < 1/2$ for $1 \leq i \leq 3$. Points in the SAW K have the property that at least two of the coordinates are integers. The points in the pushoff K_ξ are obtained by adding the vector $\xi = (\xi_1, \xi_2, \xi_3)$ to each of the points of K. Suppose now that $(x, y, z) \in K$, and that $(x+\xi_1, y+\xi_2, z+\xi_3) \in K \cap K_\xi$. By the pigeonhole principle, at least one of the following is true: both x and $x + \xi_1$ are integers; both y and $y + \xi_2$ are integers; both z and $z + \xi_3$ are integers. This means that at least one of $\{\xi_1, \xi_2, \xi_3\}$ is a non-zero integer, which is impossible. The validity of claim 1 means that $Lk(K, K_\xi)$ is defined for any pushoff in the interior of any of the 8 octants of S.

Claim (2). If ξ and ζ lie in the interior of the same octant, then $Lk(K, K_\xi) = Lk(K, K_\zeta)$. Consider the great circle on S that goes through both ξ and ζ. The points ξ and ζ separate the great circle into two arcs; the shorter of these arcs is contained in the interior of the octant containing both ξ and ζ. The points along this shorter arc define a 1-parameter family of pushoffs, starting with ξ and ending with ζ. Thus the curve K_ξ can be isotoped to K_ζ in the complement of the curve K, so we conclude that $Lk(K, K_\xi) = Lk(K, K_\zeta)$. Since the area on S of each of the octants is $\pi/8$, we conclude that $\omega(K)$ can be computed as the average of 8 directional writhes, one direction chosen from the interior of each of the 8 octants. We can however do better than this — we only need to average 4 directional writhes, since the directional writhe on any octant is the same as the directional writhe on its antipodal octant, as shown in the next claim.

Claim (3). If ξ is in the interior of an octant, then $Lk(K, K_\xi) = Lk(K, K_{-\xi})$. For the parameter t with $-1 \leq t \leq 0$, let $K_{t\xi}$ be obtained by adding the vector $t\xi$ to each point of K. As in claim 1, for each value of t, the curves $K_{t\xi}$ and K_ξ are disjoint. To see this, suppose that $(x+\xi_1, y+\xi_2, z+\xi_3) = (x'+\xi_1, y'+\xi_2, z'+\xi_3)$ for (x, y, z) and $(x', y', z') \in K$. Suppose also that both x and x' are integers. Then $(x' - x) = (1-t)\xi_1$. But $(1-t)\xi_1$ cannot be an integer because $1 \le (1-t) \le 2$ and $0 < \xi_1 < 1/2$. Therefore we can isotope K to $K_{-\xi}$ in the complement of K_ξ, so $Lk(K_\xi, K) = Lk(K_\xi, K_{-\xi})$. Likewise, we can isotop K_ξ to K in the complement of $K_{-\xi}$, so $Lk(K_{-\xi}, K_\xi) = Lk(K_{-\xi}, K)$. By symmetry of linking numbers in R, we conclude $Lk(K, K_\xi) = Lk(K, K_{-\xi})$.

$\square$

Theorem 4.2 (LS,Cim,LaS). *For any SAP K, the writhe $\omega(K)$ is the average of 4 directional writhes, with the directions chosen in any 4 mutually nonantipodal octants of S, and $4\omega(K)$ is an integer.*

Now we would like to investigate the properties of the ensemble of rooted n-SAPs. If K is an n-SAP, let K^* denote the mirror image of K. Writhe has the property that $\omega(K) = -\omega(K^*)$; so if we were average the writhe over all n-SAPs, the expected value of the writhe $\langle \omega \rangle_n = 0$ by symmetry. Consequently, we are interested in the expected value of the absolute value of the writhe $\langle |\omega| \rangle_n$, or the square of the writhe $\langle \omega^2 \rangle_n$. More generally, we are interested in the distribution of $\langle |\omega| \rangle_n$ over the set of n-SAPs.

Let $P = (0, 0, 0)$ and $Q = (0, 1, 0)$ in Z^3. Both P and Q are on the boundary of a solid cube C of size $2 \times 2 \times 2$ whose corners are:

$$\{(0, -1, 0), (0, 1, -1), (0, -1, -1), (0, 1, 1),$$
$$(2, -1, 0), (2, 1, -1), (2, -1, -1), (2, 1, 1)\}$$

The cube C is symmetric about the plane $z = 0$. Let $\{i, j, k\}$ denote unit vectors in a right-handed coordinate system for Z^3 . Consider the 10-SAP B contained in C and described as follows: begin at P, and follow the sequence of steps $\{i, i, -k, -j, -i, j, j, k, -i, -j\}$. Let $\omega(B)$ denote the writhe of B. In order to use Theorem 4.2 above, we choose pushoff directions as follows: $v_1 = (1, 1, 1)$, $v_2 = (-1, 1, 1)$, $v_3 = (-1, -1, 1)$, $v_4 = (1, -1, 1)$. Let $\{B_1, B_2, B_3, B_4\}$ denote pushoffs of B in the 4 pushoff directions $\{v_1, v_2, v_3, v_4\}$. For each pushoff direction we choose a small enough scalar to ensure that adding that scalar multiple of the pushoff direction to B to create the pushoff creates no intersections between B and its pushoff. By inspection, we have that $Lk(B, B_1) = Lk(B, B_3) = Lk(B, B_4) = +1$ and $Lk(B, B_2) = -1$. Hence $\omega(B) = +1/2$. If B^* denotes the mirror image of B (reflected in the plane $z = 0$), then $\omega(B^*) = -1/2$.

Suppose now that A is a SAP which (given one of its two orientations) intersects the cube C only in the SAW B' which begins at P and ends at Q and traverses all the steps of B except the last one: that is, B' starts at P and consists of the 9 steps $\{i, i, -k, -j. -i, j, j, k, -i\}$. We can truncate the polygon A by deleting the 9 steps of B' and adding in the step i which connects P to Q in the boundary of C, generating the new polygon A'. The polygon A is obtained by concatenating A' and B.

Lemma 4.3 (ROS). $\omega\langle A \rangle = \omega(A') + \omega(B)$.

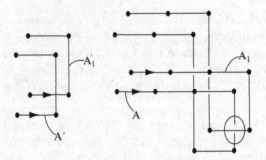

Fig. 2 [ROS]: Additivity of writhe.

Proof. Consider the following pushoffs A_1 of A and A_1' of A' in direction v_1. Fig. 2 shows the projection down the Z-axis of these curves near the cube C; the remainder of the projection of the polygon has been suppressed. In Fig. 2, the $(+)$ crossing in the circle is where A_1 crosses over A in the interior of cube C. By a small move (isototpy in the interior of C), one can move curve A straight up, crossing through curve A_1 until A now goes over A_1 in the circle, and no other crossings have been changed. This gives a pair of curves which are isotopic (by an isotopy in the interior of C) to the pair $\{A', A_1'\}$. This proves that

$$Lk(A, A_1) = Lk(A', A_1') + 1 = Lk(A', A_1') + Lk(B, B_1).$$

A similar calculation for each of the other 3 pushoff directions gives:

$$Lk(A, A_i) = Lk(A', A_i') + 1 = Lk(A', A_i') + Lk(B, B_i) \qquad i = 3, 4$$

and

$$Lk(A, A_1) = Lk(A', A_1') + 1 = Lk(A', A_1') + Lk(B, B_1) \qquad i = 2.$$

Averaging these 4 equations completes the proof of Lemma 4.3. $\square$

Theorem 4.4 (ROS). *For every function $f(n) = o(\sqrt{n})$, the probability that $\langle |\omega| \rangle_n < f(n)$ goes to zero as n goes to infinity.*

Proof. We use the Kesten Pattern Theorem [Kes] and a coin-tossing argument [DPS,ROS]. We call the $(3, 1)$ ball pair consisting of any translate of the SAW B' and the surrounding cube C a *pattern* $P = \{C, B'\}$. Let the pattern P^* be the ball pair $\{C, B'^*\}$ where B'^* is the mirror image of B' (reflected in the plane $z = 0$). Kesten's pattern theorem implies that there is a positive number ε such that for all except exponentially few sufficiently long n-SAPs, there are at least $\lfloor \varepsilon n \rfloor$ pairwise disjoint translates of C, each of which intersects the SAW in a translate of B' or B'^*. The distribution of patterns among the copies of C is analogous to tossing a coin, since each of the patterns B' and B'^* occur independently with probability $1/2$ in each of the $\lfloor \varepsilon n \rfloor$ translations of the cube C. Consequently the probability that B' occurs exactly k times among the $\lfloor \varepsilon n \rfloor$ occurrences of either B' or B'^* is less than $(1/\sqrt{\lfloor \varepsilon n \rfloor})$ for every $k \leq \lfloor \varepsilon n \rfloor$. (This can be seen by using Stirling's approximation to the binomial distribution.) The fraction of polygons with at least $\lfloor \varepsilon n \rfloor$ occurrences of either P or P^* is at least $(1 - e^{-\gamma n})$ for some positive γ. For each of these SAPs, the writhe is the sum of two terms (Lemma 4.3). The first term is from the polygon formed by truncating $\lfloor \varepsilon n \rfloor$ times, and the second term is from the $\lfloor \varepsilon n \rfloor$ copies of B or B^* formed in these truncations. If the total writhe of that SAP is less than $f(n)$, then the contribution to the writhe from the $\lfloor \varepsilon n \rfloor$ occurrences of the patterns must be one of at most $\lceil 2f(n) + 1 \rceil$ different values. Hence

$$\mathrm{Prob}(\langle |\omega| \rangle_n < f(n)) \leq \frac{(1 - e^{-\gamma n}) \lceil 2f(n) + 1 \rceil}{\sqrt{\varepsilon n}}$$

which goes to zero as n goes to infinity if $f(n) = o(\sqrt{n})$. $\qquad\square$

Theorem 4.4 strongly suggests that $\langle |\omega| \rangle \sim n^{\alpha}$. Monte Carlo simulation [ROS] for values of n between 400 and 1100 give a writhe distribution that is symmetric about the origin and sharply peaked at the origin ($n = 400$), and less sharply peaked at the origin as n increases. When values of $\log(\langle |\omega| \rangle)$ are plotted against $\log n$, the evidence [ROS] (Fig. 4) for linear behavior is excellent, and produces the following estimate: $\alpha = 0.522 \pm 0.004$.

Are almost all long randomly chosen SAP's chiral? The answer is yes. One invariant that detects chirality (inequivalence of an unoriented knot and its mirror image) is the *signature* $\sigma(K)$ of the knot K [BZ]. If K^* denotes the mirror image of K, then $\sigma(K^*) = -\sigma(K)$, so if K is *achiral* ($K = K^*$), then $\sigma(K) = 0$. The signature is additive on connected sums, so $\sigma(K \sharp L) = \sigma(K) + \sigma(L)$. One can adapt the writhe argument above with the trefoil pattern T replacing the curl pattern B, and a suitable rectangular parallelepiped replacing the cube C. Let $\langle |\sigma| \rangle_n$ denote the average of the absolute value of the signature, averaged over all n-SAPs.

Corollary 4.5 (DPS). *For every function $f(n) = o(\sqrt{n})$, the probability that $\langle |\sigma| \rangle_n < f(n)$ goes to zero as n goes to infinity.*

Hence, we conclude that the average of the absolute value of the signature of random n-SAPs grows at least as fast as $\sqrt{n}$, and that most long random knots are chiral. This result was first proved for Gaussian random polygons [DPS] by the same method.

5 Application of Random Knotting to Viral DNA Packing

Knots and links are of biological interest because they can detect and preserve topological information, especially information about DNA and the enzymes that act on DNA. Knotted DNA molecules can be well characterized experimentally by gel electrophoresis and microscopy (both transmission electron and atomic force microscopy), and therefore used as assays for different biochemical reactions. Characterization of knotted products formed by random cyclization of linear molecules has been used to quantify important biochemical properties of DNA such as its effective diameter [RCV,ShW]. DNA knots and catenanes obtained as the product of site-specific recombination have also been a key to unveiling the mechanism of enzymatic action [WC,SBS]. In both cases, the development of mathematical and computational tools has greatly enhanced analysis of the experimental results [FLA,ES]

Significant numbers of DNA knots are found also in biological systems: in *Escherichia coli* cells harboring mutations in the *GyrB* or *GyrA* genes [SKI], bacteriophages P2 and P4 [LDC,LPC], and cauliflower mosaic viruses [MML]. However, very little biological information about these systems has been inferred from the observed knots. In particular, interpretation of the experimental results for bacteriophages has been limited by the experimental difficulty in quantifying the complex spectrum of knotted products. These difficulties have paralleled those encountered in developing a theory for random knotting of ideal polymeric chains in cases where interactions with other macromolecules and/or confinement in small volumes have a significant function [Man,TRO,MMO].

Bacteriophages are viruses that infect bacteria. They pack their double-stranded DNA genomes to near-crystalline density in viral capsids and achieve one of the highest levels of DNA condensation found in nature. When I was on sabbatical in Berkeley in 1989, Jim Wang described to me the problem of DNA packing in icosehedral viral capsids, and the high degree of knotting produced when the viral DNA is released from the capsids. Despite numerous studies some essential properties of the packaging geometry of the DNA inside the phage capsid are still unknown. Although viral DNA is linear double-stranded with sticky (cohesive) ends, the linear viral DNA quickly becomes cyclic when removed from the capsid, and for some viral DNA the observed knot probability is an astounding 95%. In the summer of 1998, my PhD students Javier Arsuaga and Mariel Vazquez spent 2 months in the laboratory of Joaquim Roca in Barcelona, supported by the Burroughs Wellcome Interfaces grant to the Program in Mathematics and Molecular Biology. In the Roca laboratory, they infected bacterial stock, harvested viral capsids and extracted and analyzed the viral DNA. They quantified the DNA knot spectrum produced in the experiment, and used Monte Carlo generation of knots in confined volumes to compare a random knot spectrum to the observed viral DNA knot spectrum. A series of papers were produced as a result of this collaboration [TAV,AVT,ArT,AVM], and I will describe some of the results from these papers, focusing on (and reproducing here) most of the discussion and analysis from the most recent PNAS paper [AVM].

All icosahedral bacteriophages with double-stranded DNA genomes are believed to pack their chromosomes in a similar manner [EC]. During phage morphogenesis, a procapsid is first assembled, and a linear DNA molecule is actively introduced inside it by the connector complex [RHA,STS]. At the end of this process, the DNA and its associated water molecules fill the entire capsid volume, where DNA reaches concentrations of 800 mg/ml [KCS]. Some animal viruses [SB] and lipoDNA complexes used in gene therapy [SDD] are postulated to hold similar DNA arrangements as those found in bacteriophages. Although numerous studies have investigated the DNA packing geometry inside phage capsids, some of its properties remain unknown. Biochemical and structural analyses have revealed that DNA is kept in its B form [ACT,EH,LDB] and that there are no specific DNAprotein interactions [HMC,Ser] or correlation between DNA sequences and their spatial location inside the capsid, with the exception of the cos ends in some viruses. Many studies have found that regions of the packed DNA form domains of parallel fibers, which in some cases have different orientations, suggesting a certain degree of randomness [EH,LDB]. The above observations have led to the proposal of several long-range organization models for DNA inside phage capsids: the ball of string model [RWC], the coaxial spooling model [EH,CCR,RWC,CCR], the spiral-fold model [BNB], and the folded toroidal model [Hud]. Liquid crystalline models, which take into account properties of DNA at high concentrations and imply less global organization, have also been proposed [LDB]. Cryo EM and spatial symmetry averaging has recently been used to investigate the surface layers of DNA packing [JCJ]. In [AVM], the viral DNA knot spectrum is used to investigate the packing geometry of DNA inside phage capsids.

The bacteriophage P4 has a linear, double-stranded DNA genome that is 10–11.5 kb in length and flanked by 16-bp cohesive cos ends [WMC]. It has long been known that extraction of DNA from P4 phage heads results in a large proportion of highly knotted, nicked DNA circles [LDC,LPC]. DNA knotting probability is enhanced in P4 derivatives containing genome deletions [WMH] and in tailless mutants [IJC]. Most DNA molecules extracted from P4 phages are circles that result from the cohesive-end joining of the viral genome. Previous studies have shown that such circles have a knotting probability of about 20% when DNA is extracted from mature P4 phages [LDC]]. This high value is increased more than 4-fold when DNA is extracted from incomplete P4 phage particles (which we refer to as capsids) or from noninfective P4 mutants that lack the phage tail (which we refer to as tailless mutants [LDC]). Knotting of DNA in P4 deletion mutants is even greater. The larger the P4 genome deletion, the higher the knotting probability [WMH]. For P4 vir1 del22, containing P4s largest known deletion (1.6 kb deleted [RDM]), knotting probability is more than 80% [IJC]. These values contrast with the knotting probability of 3% (all trefoil knots) observed when identical P4 DNA molecules undergo cyclization in dilute free solution [RCV,RUV]. These differences are still more striking when the variance in distribution of knot complexity is included. Although knots formed by random cyclization of 10-kb linear DNA in free solution have an average crossing number of

three [RCV,RUV], knots from phage particles have a knotting probability of
95% and appear to have very large crossing numbers, averaging about 26.
[LDC,WMH,IJC].

The reasons for the high knotting probability and knot complexity of bacteriophage DNA have been investigated. Experimental measurements of the
knotting probability and distribution of knotted molecules for P4 vir1 del22
mature phages, capsids, and tailless mutants was performed by 1- and 2-
dimensional gel electrophoresis, followed by densitometer analysis. We will
describe the Monte Carlo simulations to determine the effects that the confinement of DNA molecules inside small volumes have on knotting probability
and complexity. We conclude from our results that for tailless mutants a significant amount of DNA knots must be formed before the disruption of the
phage particle, with both increased knotting probability and knot complexity
driven by confinement of the DNA inside the capsid.

In [AVM] it is shown that the DNA knots provide information about the
global arrangement of the viral DNA inside the capsid. The distribution of
the viral DNA knots is analyzed by high-resolution gel electrophoresis. Monte
Carlo computer simulations of random knotting for freely jointed polygons
confined to spherical volumes is performed. The knot distribution produced
by simulation is compared to the observed experimental DNA knot spectrum.
The simulations indicate that the experimentally observed scarcity of the
achiral knot 4_1 and the predominance of the torus knot 5_1 over the twist
knot 5_2 are not caused by confinement alone but must include writhe bias in
the packing geometry. Our results indicate that the packaging geometry of
the DNA inside the viral capsid is non-random and writhe-directed.

5.1 Knot Type Probabilities for P4 DNA in Free Solution

The probability that a DNA knot K of n statistical lengths and diameter d is formed by random closure in free solution is given by $P_K(n,d) =
P_K(n,0)e^{-rd/n}$, where r depends on the knot type and equals 22 for the
trefoil knot 3_1 and 31 for the figure 8 knot 4_1 [RCV,ShW]. The knotting
probability of a 10-kb DNA molecule cyclized in free solution is 0.03, which
implies an effective DNA diameter near 35 Å. Because $P_{3_1}(34,0) = 0.06$ and
$P_{4_1}(34,0) = 0.009$, then $P_{3_1}(34,35) = 0.027$ (1/36 times that of the unknot)
and $P_{4_1}(34,35) = 0.003$ (1/323 times that of the unknot). These values were
used to estimate the fractions of the knot 3_1 and the knot 4_1 generated for
P4 DNA in free solution.

5.2 Monte Carlo Simulation

Knotting probabilities of equilateral polygons confined into spherical volumes were calculated by means of Markov-chain Monte Carlo simulations followed by rejection criteria. Freely jointed closed chains, composed of n equilateral segments, were confined inside spheres of fixed radius, r, and sampled: values of n ranged from 14 to 200 segments; r values, measured as multiples of the polygonal edge length, ranged from 2 to infinity. Excluded volume effects were not taken into account. Markov chains were generated by using the Metropolis algorithm [MRR]. The temperature, a computational parameter, was held at $T = 300$ K to improve the efficiency of the sampling algorithm. Other values of T produced similar results, thus indicating that the computation is robust with respect to this parameter. Chains contained inside the sphere were assigned zero energy. Chains lying partly or totally outside the confining sphere were assigned an energy given by the maximum of the distances of the vertices of the chain to the origin. Only chains with zero energy were sampled. A random ensemble of polygons was generated by the crankshaft algorithm as follows: (i) two vertices of the chain were selected at random, dividing the polygon into two subchains, and (ii) one of the two subchains was selected at random (with equal probabilities for each subchain), and the selected subchain rotated through a random angle around the axis connecting the two vertices. This algorithm is known to generate an ergodic Markov chain in the set of polygons of fixed length [Mil]. Correlation along the subchains was computed by using time-series analysis methods as described by Madras and Slade [MS]. Identification of the knotted polygons was achieved by computing the Alexander polynomial $\Delta(t)$ [BZ,Rol,Adm] evaluated at $t = -1$. It is known that $\Delta(-1)$ does not identify all knotted chains; however, for polygonal chains not confined to a spherical volume, nontrivial knots with trivial $\Delta(-1)$ values rarely occur. This circumstance has been observed by using knot invariants, such as the HOMFLY polynomial, that distinguish between knotted and unknotted chains with higher accuracy than the Alexander polynomial. Computer simulations for small polygons (< 55 segments) show that the knotting probabilities obtained by using $\Delta(-1)$ agree with those obtained by using the HOMFLY polynomial. Furthermore, Deguchi and Tsurusaki have reported that the value of $\Delta(-1)$ can almost always determine whether a given Gaussian polygon is unknotted for lengths ranging from 30 to 2,400 segments [DT2]. Each selected knotted polygon was further identified by evaluating its Alexander polynomial at $t = -2$ and $t = -3$. Although the Alexander polynomial is an excellent discriminator among knots of low crossing number and its computation is fast, it does not distinguish completely among some knotted chains [for example, composite knots $3_1\sharp3_1$ and $3_1\sharp4_1$ have polynomials identical to those of prime knots 8_{20} and 8_{21}, respectively [RW]. Evaluation of the polynomial at $t = -2$ and $t = -3$ is also ambiguous because the Alexander polynomial is defined up to units (power or t) in $Z[t^{-1}, t]$. To deal with this uncertainty, we followed van Rensburg and Whittington [RW] and chose the largest exponent k such that the product $n^{\pm k}\Delta(-n)$ is an odd integer with $n = 2$ or 3. This value was taken as the knot invariant. To compute the writhe, we generated > 300 regular projections and resulting knot diagrams for each selected polygon. To each of the projected crossings a sign was assigned by the skew lines convention (Fig. 1). The directional writhe for each diagram was computed by summing these values. The writhe was then determined by averaging the directional writhe over a large number of randomly chosen projections. To generate writhe-directed random distributions of polygons, we used a rejection method in which polygons whose writhe was below a positive value were not sampled.

5.3 Results and Discussion Knot Complexity of DNA Molecules Extracted from Phage P4

We extracted the 10-kb DNA from the tailless mutant of phage P4 vir1 del22, which produces 95% knotted molecules [AVT], and analyzed it by a high-resolution two-dimensional gel electrophoresis [TAV] (Fig. 3A). This technique allowed us to separate DNA knot populations according to their crossing number (i.e., the minimal number of crossings over all projections of a knot), as well as to separate some knot populations of the same crossing number [SKB,VCL]. In the first dimension (at low voltage), individual gel bands corresponding to knot populations having crossing numbers between three and nine were discernible; knots with higher crossing numbers were embedded in a long tail (denoted as K in Fig. 3 A). The second dimension (at high voltage) further resolved individual gel bands corresponding to knot populations with crossing numbers between six and nine. Although knot populations containing three, four, and five crossings (denoted as 3–5 in Fig. 3B) migrated as single bands in a main arch of low gel velocity, knot populations containing six and more crossings split into two subpopulations (denoted as 6–9 and 6'–9' in Fig. 3B), creating a second arch of greater gel velocity. Fig. 3(A) shows DNA was extracted from tailless mutants of phage P4 vir1 del22 and analyzed by two-dimensional agarose gel electrophoresis. The first dimension at low voltage (top to bottom) separated DNA knot populations according to their crossing number. The unknotted DNA circle or trivial knot (0) has the slowest gel velocity, whereas knotted DNA populations (K) have gel velocity proportional to their crossing number. The second dimension at high voltage (left to right) segregated the linear DNA molecules (L) from the arched distribution of knotted molecules and further resolved some gel bands corresponding to knot subpopulations. (B) Upper area of the gel picture showing knot populations of low crossing number. Individual gel bands corresponding to knot populations containing three to nine crossings are indicated (labeled 3–9) in the main arch of the gel. A second arch of higher

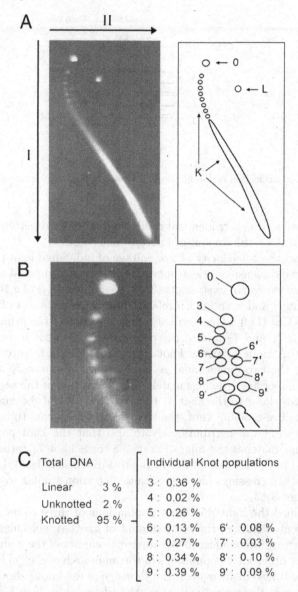

Fig. 3 [AVM]: Analysis of knotted DNA by gel electrophoresis.

gel speed containing knot subpopulations of six and more crossings is generated by the second dimension of the electrophoresis. Individual gel bands of knot subpopulations of six to nine crossings (labeled 6'–9') are indicated. (*C*) Quantification of the individual knot populations of six to nine crossings (3–9 and 6'–9'). Both densitometric and phosphorimaging reading of three independent samples of DNA extracted from tailless mutants of phage P4

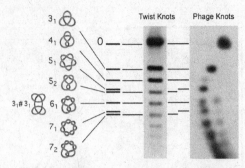

Fig. 4 [AVM]: Identification of specific knot types by their position in the gel.

vir1 del22 produced nearly identical results. The indicated percentage values are relative to the total amount of knotted molecules.

Fig. 4 shows the gel velocity at low voltage of individual knot populations resolved by two-dimensional electrophoresis (Right) is compared with the gel velocity at low voltage of twist knots (3_1, 4_1, 5_2, 6_1, and 7_2) of a 10-kb nicked plasmid (Center) and with known relative migration distances of some knot types [SKB,VCL] (Left). Geometrical representations of the prime knots 3_1, 4_1, 5_1, 5_2, 6_1, 7_1 and 7_2 and of the composite knot $3_1\sharp3_1$ are shown. The unknotted DNA circle or trivial knot (0) is also indicated. Note that in the main arch of the two-dimensional gel and below the knots 3_1 and 4_1, the knot population of five crossings matches the migration of the torus knot 5_1, which migrates closer to the knot 4_1 than to the knots of six crossings. The other possible five-crossing knot, the twist knot 5_2, appears to be negligible or absent in the viral distribution. Note also that the knot population of seven crossings matches the migration of the torus knot 7_1 rather than the twist knot 7_2. In the secondary arch of the two-dimensional gel, the first knot population of six crossings has low-voltage migration similar to that of the composite knot $3_1\sharp3_1$.

We quantified the individual knot populations of three to nine crossings, which represented 2.2% of the total amount of knotted molecules (Fig. 4C). Densitometer readings confirmed the apparent scarcity of the figure 8 knot 4_1 relative to the other knot populations in the main arch of the gel (denoted by 4 in Fig. 4B). It also made evident the shortage of the knot subpopulation of seven crossings in the second arch of the gel (denoted by $7'$ in Fig. 3B). The scarcity of the knot 4_1 relative to the knot 3_1 and to other knot populations is enhanced if we make the correction for DNA molecules plausibly knotted outside the viral capsid. Namely, if a fraction of the observed knots were formed by random cyclization of DNA outside the capsid, then, in the worst-case scenario, all observed unknotted molecules (no more than 5% of the total molecules extracted) would be formed in free solution. In such a case, one can predict that 38% of the total number of observed 3_1 knots and 75% of the observed 4_1 knots are formed by random knotting in free solution [RCV,SW].

If all of the knots plausibly formed outside the capsid were removed from the observed knot distribution, the experimental values for knots 4_1 and 3_1 (1:18 ratio) would be corrected, resulting in a 1:44 ratio.

5.4 Identification of Specific Knot Types by Their Location on the Gel

Gel electrophoresis can distinguish some knot types with the same crossing number. For example, at low voltage, torus knots (such as 5_1 and 7_1) migrate slightly slower than their corresponding twist knots (5_2 and 7_2) [SKB,VCL]. We used this knowledge in conjunction with a marker ladder for twist knots (3_1, 4_1, 5_2, 6_1, and 7_2) to identify several gel bands of the phage DNA matching the migration of known knot types (Fig. 4). In the main arch of the gel, in addition to the unambiguous knots 3_1 and 4_1, the knot population of five crossings matched the migration of the torus knot 5_1. The other possible five-crossing knot, the twist knot 5_2 that migrates between and equidistant to the four-and six-crossing knot populations, appeared to be negligible or absent. The knot population of seven crossings matched the migration of the torus knot 7_1 rather than the twist knot 7_2, which has slightly higher gel velocity. Yet, we cannot identify this gel band as the knot 7_1, because other possible knot types of seven crossings cannot be excluded.

Several indicators led us to believe that the second arch of the gel consists of mainly composite knots. First, the arch starts at knot populations containing six crossings, and no composite knots of fewer than six crossings exist. Second, the population of six crossings matched the migration at low voltage of the granny knot $3_1 \sharp 3_1$ [KKS], although the square knot $3_1 \sharp - 3_1$ cannot be excluded. Third, consistent with the low amount of 4_1 knots, the size of the seven-crossing subpopulation is also reduced: any composite seven-crossing knot is either $3_1 \sharp 4_1$ or $-3_1 \sharp 4_1$. The increased gel velocity at high voltage (second gel dimension) of composite knots relative to prime knots of the same crossing number likely reflects distinct flexibility properties of the composites during electrophoresis [WSR].

5.5 Monte Carlo Simulations of Random Knot Distributions in Confined Volumes

Next, we asked whether the observed distribution of DNA knots could be compatible with a random embedding of the DNA inside the phage capsid. We used Monte Carlo simulations to model knotting of randomly embedded, freely jointed polygons confined to spherical volumes. Because the persistence length of the duplex DNA is not applicable in confined volumes (it is

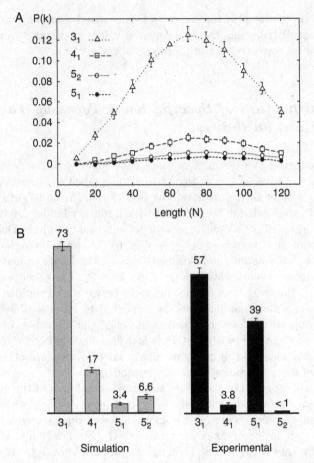

Fig. 5 [AVM]: Comparison of experimental and computer-simulated distributions of knots.

applicable in unbounded three-dimensional space), we considered freely jointed polygons as the zeroth approximation of the packed DNA molecule. Then, the flexibility of the chain is given by the ratio R/N, where N is the number of edges in the polygon and R is the sphere radius in edge-length units. When we computed random knot distributions for a range of chain lengths confined to spheres with a fixed radius, the probabilities of the knots 3_1, 4_1, 5_1, and 5_2 produced nonintersecting distributions, with simpler knots being more probable (Fig. 5A). That is, the knot 3_1 is more probable than the knot 4_1, and both are more probable than any five-crossing knot. In addition, the probability of the twist knot 5_2 is higher than that of the torus knot 5_1. Similar results had been observed for other random polymer models with/without volume exclusion and with/without confinement [DT2,Man], indicating that this phenomenon is model-independent. All of the simulated distributions, showing the monotonically decreasing amounts of knotted products with increasing crossing number, highly contrasted with our experimental distribution, in which the probability of the knot 4_1 is markedly reduced and in which the knot probability of the knot 5_1 prevails over that of the knot 5_2 (Fig. 5B). These differences provide a compelling proof that the embedding of the DNA molecule inside the phage capsid is not random.

Fig. 5(A) shows the distribution probabilities $P(k)$ obtained by Monte Carlo simulations of the prime knots 3_1, 4_1, 5_1, and 5_2 for closed ideal polymers of variable chain lengths (n = number of edges) confined to a spherical volume of fixed radius ($R = 4$ edge lengths). Error bars represent standard deviations. (B) Comparison of the computed probabilities of the knots 3_1, 4_1, 5_1, and 5_2 (for polymers of length $n = 90$ randomly embedded into a sphere of radius $R = 4$) with the experimental distribution of knots. The relative amount of each knot type is plotted. Note that fractions of knots 3_1 and 4_1 plausibly formed in free solution are not subtracted from the experimental distribution. If these corrections are considered, the relative amount of knot 4_1 is further reduced.

Fig. 6 shows the writhe of polygons of length $n = 90$ randomly embedded into a sphere of radius $R = 4$ were computed, and only conformations whose writhe values were higher than a fixed value ($Wr = 4$, 6, or 8) were sampled. The computed mean writhe value ($\langle Wr \rangle$) of each sampled population is indicated. The ratios of the probabilities of the knots 4_1, 5_1, and 5_2 relative to that of the knot 3_1 for each writhe-biased sampling are plotted (P).

How can we explain the scarcity of 4_1 in the spectrum of viral knots? The knot 4_1 is achiral (equivalent to its mirror image). Random polygonal realizations of the 4_1 knot in free space and in confined volumes produce a family of polygons whose writhe distribution for any polygonal length is a Gaussian

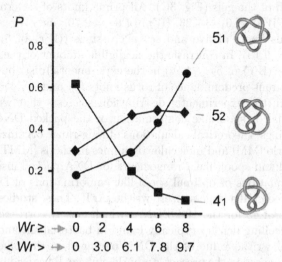

Fig. 6 [AVM]: Effect of a writhe-biased sampling on the probability of knots 4_1, 5_1, and 5_2.

curve with zero mean (the writhe is a geometrical quantity measuring the signed spatial deviation from planarity of a closed curve) and whose variance grows as the square root of the length (Theorem 7). Therefore, we argue that the main reason for the scarcity of the knot 4_1 is a writhe bias imposed on the DNA inside the phage capsid. To test this hypothesis, we simulated polygons randomly embedded in spheres whose mean writhe value was gradually increased. To induce writhe in the sampling, we used a rejection method in which polygons of writhe below a cutoff value were not sampled. Then, we calculated the probabilities of the prime knots 4_1, 5_1, and 5_1 for each writhe-biased sampling. The results shown in Fig. 6 were computed with a freely jointed chain of 90 edges confined in a sphere of radius of 4 edge-length units. A drop of the probability of the knot 4_1, as well as an exponential increase of the probability of the torus knot 5_1 but not of the twist knot 5_2, readily emerged by increasing the writhe rejection value. The same results, but with knots of opposite sign, were obtained for knot distributions with the corresponding negative writhe values. These writhe-induced changes in the knot probability distribution are independent of the number of edges in the equilateral polygon and the sphere radius length. Accordingly, previous studies had shown that the mean writhe value of random conformations of a given knot does not depend on the length of the chain but only on the knot type and that these values are model-independent [RSW]. Because the writhe-directed simulated distributions approach the observed experimental spectrum of knots, we conclude that a high writhe of the DNA inside the phage is the most likely factor responsible for the observed experimental knot spectrum.

Consistent with the involvement of writhe in the DNA packing geometry, it is also the reduced amount of prime knots of six crossings visible in the main arch of the gels (Fig. 3C). All prime knots of six crossings have a lower $\langle Wr \rangle$ ($\langle Wr \rangle$ of $6_1 = 1.23$, $\langle Wr \rangle$ of $6_2 = 2.70$, and $\langle Wr \rangle$ of $6_3 = 0.16$) than the torus knots of five and seven crossings ($\langle Wr \rangle$ of $5_1 = 6.26$ and ($\langle Wr \rangle$ of $7_1 = 9.15$). In contrast, the negligible amount of the twist knot of five crossings ($\langle Wr \rangle$ of $5_2 = 4.54$) in the experimental distributions is striking. The apparent predominance of torus knots (5_1 and 7_2) over twist knots (5_2 and 7_2) in the experimental distribution suggests that writhe emerges from a toroidal or spool-like conformation of the packed DNA. Consistent with our findings, theoretical calculations of long-range organization of DNA by Monte Carlo [MM] and molecular dynamics methods [MMT,ArT,TK,LL] favor toroidal and spool-like arrangements for DNA packed inside the phage capsids. Calculations of optimal spool-like conformations of DNA in phage P4 already predicted a large nonzero writhe [ArT]. These studies gave an estimated writhe of 45 for the 10-kb DNA, which closely corresponds with the level of supercoiling density typically found in bacterial chromosomes [ArT].

The actual writhe value of the DNA packaged in the phage P4 capsid cannot be estimated in the present study. The phage P4 capsid has a diameter of 38 nm. If the parameters used to compute writhe-biased ensembles as in

Fig. 6 ($n = 90$ and $R = 4$) were applied to a 10-kb DNA molecule, they would translate into 90 segments of 35 nm confined in a model capsid of radius 140 nm. Likewise, our study cannot argue for or against recent models that suggest that to minimize DNA bending energy, a spool conformation might be concentric rather than coaxial [LL]. Therefore, beyond the main conclusion of this work that the distribution of viral knots requires the mean writhe of the confined DNA be nonzero, the applicability of our simulations to other aspects of the DNA packaging in phage P4 is limited. We argue that further identification of the knotted DNA populations will provide more critical information for the packing geometry of DNA inside the phage.

Knots can be seen as discrete measuring units of the organizational complexity of filaments and fibers. Here, we show that knot distributions of DNA molecules can provide information on the long-range organization of DNA in a biological structure. We chose the problem of DNA packing in an icosahedral phage capsid and addressed the questions of randomness and chirality by comparing experimental knot distributions with simulated knot distributions. The scarcity of the achiral knot 4_1 and the predominance of the torus knot 5_1 in the experimental distribution highly contrasted with simulated distributions of random knots in confined volumes, in which the knot 4_1 is more probable than any five-crossing knot, and the knot 5_2 is more probable the knot 5_1. To our knowledge, these results produce the first topological proof of nonrandom packaging of DNA inside a phage capsid. Our simulations also show that a reduction of the knot 4_1 cannot be obtained by confinement alone but must include writhe bias in the conformation sampling. Moreover, in contrast to the knot 5_2, the probability of the torus knot 5_1 rapidly increases in a writhe-biased sampling. Given that there is no evidence for any other biological factor that could introduce all of the above deviations from randomness, we conclude that a high writhe of the DNA inside the phage capsid is responsible for the observed knot spectrum and that the cyclization reaction captures that information.

References

[Ax1]   Alexander, J.W.: An Example of a Simply Connected Surface Bounding a Region which is not Simply Connected. Proceedings of the National Academy of Sciences USA, **10**, 8–10 (1924).

[Ax2]   Alexander, J.W.: On the Subdivision of 3-Space by a Polyhedron. Proceedings of the National Academy of Sciences USA, **10**, 6–8 (1924).

[Sv]   Silver, D.S.: Knot Theory's Odd Origins. American Scientist, **94**, 58–66 (2006).

[Mof]   Moffatt, H.K.: Knots and Fluid Dynamics. In Ideal Knots, Series on Knots and Everything, eds. Stasiak, A., Katritch, V., Kauffman, L.H. (World Scientific, Singapore), Vol. **19**, 223–233 (1999).

[Ric1]   Ricca, R.L.: New Developments in Topological Fluid Mechanics: From Kelvin's Vortex Knots to Magnetic Knots. In Ideal Knots, Series on Knots and Everything, eds. Stasiak, A., Katritch, V., Kauffman, L.H. (World Scientific, Singapore), Vol. **19**, 255–273 (1999).

[Ric2]   Ricca, R.L.: Tropicity and Complexity Measures for Vortex Tangles, Lecture Notes in Physics v. 571. Springer, Berlin Heidelberg New York, 366–372 (2001).

[LZC]   Liu, Z, Zechiedrich, E.L., Chan, H.S.: Inferring global topology from local juxta-position geometry: interlinking polymer rings and ramifications for topoisomerase action. Biophys. J., **90**, 2344–2355 (2006).

[BZ]   Buck, G, Zechiedrich, E.L.: DNA disentangling by type-2 topoisomerases. J. Mol Biol. Jul 23; 340(5): 933–9 (2004).

[FW]   Frisch, H.L., Wasserman, E.: Chemical Topology. J. Am. Chem. Soc. **83**, 3789–3795 (1961).

[Was]   Wasserman, E.: Chemical Topology. Scientific American **207**, 94–102 (1962)

[Del]   Delbruck, M: in Mathematical Problems in the Biological Sciences, Proceedings of Symposia in Applied Mathematics **14**, 327– (1962).

[VLF]   Vologodskii, A.V., Lukashin, A.V., Frank-Kemenetkii, M.D., Anshelevich, V.V.: The Knot Problem in Statistical Mechanics of Polymer Chains. Sov. Phys.-JETP **39**, 1059– (1974).

[CM]   des Cloizeaux, J., Mehta, M.L.: J. de Physique **40**, 665– (1979).

[MW]   Michels, J.P.J., Wiegel, F.W.: Probability of Knots in a Polymer Ring. Phys. Lett. **90A**, 381–384 (1984).

[SW]   Sumners, D.W., Whittington, S.G.: Knots in Self-Avoiding Walks. J. Phys. A: Math. Gen. **21**, 1689–1694 (1988).

[Kes]   Kesten, H.: On the Number of Self-Avoiding Walks. J. Math. Phys. **4**, 960–969 (1963).

[Pip]   Pippenger, N.: Knots in Random Walks. Discrete Appl. Math. **25**, 273-278 (1989).

[Adm]   Adams, C.C.: The Knot Book. W.H. Freeman and Co., New York (1991).

[RSW]   van Rensburg, E.J.J., Sumners, D.W., Wasserman, E., Whittington, S.G.: Entan-glement Complexity of Self-Avoiding Walks. J. Phys. A.: Math. Gen. **25**, 6557–6566 (1992).

[RBHS]   Lacher R.C., Bryant J.L., Howard L., Sumners D.W. Linking phenomena in the amorphous phase of semicrystalline polymers. Macromolecules **19**, 2639–2643 (1986).

[DNS]   Diao, Y.,Nardo, J.C., Sun Y.: Global Knotting in Equilateral Polygons. J. of Knot Theory and Its Ramifications **10**, 597–607 (2001).

[Do2]   Diao, Y.: The Knotting of Equilateral Polygons in R^3. J. of Knot Theory and Its Ramifications **4**, 189–196 (1995).

[DPS]   Diao, Y., Pippenger, N., Sumners, D.W.: On Random Knots. J. of Knot Theory and Its Ramifications **3**, 419–424 (1994).

[Do1],   Diao, Y.: Minimal Knotted Polygons on the Simple Cubic Lattice. J. of Knot Theory and Its Ramifications **2**, 413–425 (1993).

[RW]   van Rensburg, E.J.J., Whittington, S.G.: The Knot Probability in Lattice Polygons. J. Phys. A.: Math. Gen. **23**, 3573–3590 (1990).

[SSW]   Soteros C., Sumners D.W., Whittington S.G.: Entanglement complexity of graphs in Z^3, Math. Proc Camb. Phil. Soc. **111**, 75–91 (1992).

[Jun]   Jungreis, D.: Gaussian Random Polygons are Globally Knotted. J. of Knot Theory and Its Ramifications **4**, 455–464 (1994).

[Ken]   Kendall:The knotting of brownian motion in 3-space. J. Lon. Math. Soc. **19**, 378– (1979).

[DT1]   Deguchi, T., Tsurasaki, K.: Universality of Random Knotting. Phys. Rev. E. **55**, 6245–6248 (1997).

[Nak]   Nakanishi, Y.: A Note on Unknotting Number. Math. Sem. Notes Kobe Univ. **9**, 99–108 (1981).

[Ful]   Fuller, B.: The Writhing Number of a Space Curve. Proc. Nat. Acad. Sci. USA **68**, 815–819 (1971).C

[LS]   Lacher, R.C., Sumners D.W.: Data structures and algorithms for computation
          of topological invariants of entanglements: link, twist and writhe. In Computer
          Simulation of Polymers, Prentice-Hall, Roe, R.J., ed. Englewood Cliffs, NJ, 365–
          373 (1991).
[Cim]   Cimasoni, D.: Computing the Writhe of a Knot. J. Knot Theory and Its
          Ramifications 10, 387–395 (2001).
[LaS]   Laing, C, Sumners D.W.: Computing the Writhe on Lattices. J. Phys A, Math.
          Gen. 39, 3535–3543 (2006).
[ROS]   van Rensburg, E.J.J., Orlandini, E., Sumners, D.W., Tesi. M.C., Whittington,
          S.G.: The writhe of a self-avoiding polygon. J. Phys. A Math. Gen. 26, L981–L986
          (1993).
[BZ2]   Burde, G., Zieschang, H.: Knots. De Gruyter, Berlin, New York (1985).
[Rol]   Rolfsen, D.: Knots and Links. Publish or Perish (1976).
[RCV]   Rybenkov, V. V., Cozzarelli, N. R. & Vologodskii, A. V.: Probability of DNA
          Knotting and the Effective Diameter of the DNA Double Helix. Proc. Natl. Acad.
          Sci. USA 90, 5307–5311 (1993).
[ShW]   Shaw, S. Y., Wang, J. C.: Knotting of a DNA chain during ring closure. Science
          260, 533–536 (1993).
[WC]   Wasserman, S. A., Cozzarelli, N. R.: Biochemical topology: applications to DNA
          recombination and replication. Science 232, 951–960 (1986).
[SBS]   Stark, W.M., Boocock, M.R., Sherratt, D. J.: Site-specific recombination by Tn3
          resolvase. Trends Genet. 5, 304–309 (1989).
[FLA]   Frank-Kamenetskii, M.D., Lukashin, A.V., Anshelevich, V.V., Vologodskii, A.V.:
          Torsional and bending rigidity of the double helix from data on small DNA rings.
          J. Biomol. Struct. Dyn. 2, 1005–1012 (1985).
[ES]   Ernst, C., Sumners, D.W.: A Calculus for Rational Tangles: Applications to DNA
          Recombination. Math. Proc. Camb. Phil. Soc. 108, 489–515 (1990).
[SKI]   Shishido, K., Komiyama, N., Ikawa, S.: Increased production of a knotted form of
          plasmid pBR322 DNA in Escherichia coli DNA topoisomerase mutants. J. Mol.
          Biol. 195, 215–218 (1987).
[LDC]   Liu, L. F., Davis, J. L., Calendar, R.: Novel topologically knotted DNA from
          bacteriophage P4 capsids: studies with DNA topoisomerases. Nucleic Acids Res.
          9, 3979–3989 (1981).
[LPC]   Liu, L. F., Perkocha, L., Calendar, R.. Wang, J. C.: Knotted DNA from
          Bacteriophage Capsids. Proc. Natl. Acad. Sci. USA 78, 5498–5502 (1981).
[MML]   Menissier, J., de Murcia, G., Lebeurier, G., Hirth, L.: Electron microscopic studies
          of the different topological forms of the cauliflower mosaic virus DNA: knotted
          encapsidated DNA and nuclear minichromosome. EMBO J. 2, 1067–1071 (1983).
[Man]   Mansfield, M. L.: Knots in Hamilton Cycles. Macromolecules 27, 5924–5926
          (1994).
[TRO]   Tesi, M. C., van Resburg, J.J.E., Orlandini, E., Whittington, S. G.: Knot prob-
          ability for lattice polygons in confined geometries. J. Phys. A: Math. Gen 27,
          347–360 (1994).
[MMO]   C. Micheletti, C, Marenduzzo, D., Orlandini, E., Sumners, D.W.: Knotting of
          Random Ring Polymers in Confined Spaces. J. Chem. Phys. 124, 064903 (2006).
[TAV]   Trigueros, S, Arsuaga, J., Vazquez, M.E., Sumners, D.W., Roca, J.: Novel display
          of knotted DNA molecules by two-dimensional gel electrophoresis. Nucleic Acids
          Research 29, 67–71 (2001).
[AVT]   J. Arsuaga, M. Vazquez, S. Trigueros, D.W. Sumners and J. Roca, Knotting
          probability of DNA molecules confined in restricted volumes: DNA knotting in
          phage capsids, Proc. National Academy of Sciences USA 99, 5373–5377 (2002).
[ArT]   Arsuaga, J., R. Tan, K-Z, Vazquez, M.E., Sumners, D.W., Harvey, S.C.: Inves-
          tigation of viral DNA packing using molecular mechanics models. Biophysical
          Chemistry 101–102, 475–484 (2002).

[AVM]   Arsuaga, J., Vazquez, M.E., McGuirk, P., Sumners, D.W., Roca, J.: DNA Knots
         Reveal Chiral Organization of DNA in Phage Capsids. Proc. National Academy
         of Sciences USA **102**, 9165–9169 (2005).

[EC]   Earnshaw, W. C., Casjens, S. R.: DNA packaging by the double-stranded DNA
         bacteriophages. Cell **21**, 319–331 (1980).

[RHA]   Rishov, S., Holzenburg, A., Johansen, B.V., Lindqvist, B.H.: Bacteriophage P2
         and P4 Morphogenesis: Structure and Function of the Connector. Virology **245**,
         11–17 (1998).

[STS]   Smith, D.E., Tans, S.J., Smith, S.B., Grimes S., Anderson, D.L., Bustamante, C.:
         The bacteriophage φ29 portal motor can package DNA against a large internal
         force. Nature **413**, 748–752 (2001).

[KCS]   Kellenberger, E., Carlemalm, E., Sechaud, J., Ryter, A., Haller, G.: Considera-
         tions on the condensation and the degree of compactness in non-eukaryotic DNA:
         in Bacterial Chromatin, eds. Gualerzi, C. & Pon, C. L. (Springer, Berlin), p. 11
         (1986).

[SB]   San Martin, C., Burnett, R.: Structural studies on adenoviruses. Curr. Top.
         Microbiol. Immunol. **272**, 57–94 (2003).

[SDD]   Schmutz, M., Durand, D., Debin, A., Palvadeau, Y., Eitienne, E.R., Thierry, A.
         R.: DNA packing in stable lipid complexes designed for gene transfer imitates
         DNA compaction in bacteriophage. Proc. Natl. Acad. Sci. USA **96**, 12293–12298
         (1999).

[ACT]   Aubrey, K., Casjens, S., Thomas, G.: Secondary structure and interactions of the
         packaged dsDNA genome of bacteriophage P22 investigated by Raman difference
         spectroscopy. Biochemistry **31**, 11835–11842 (1992).

[EH]   Earnshaw, W. C., Harrison, S.: DNA arrangement in isometric phage heads.
         Nature **268**, 598–602 (1977).

[LDB]   Lepault, J., Dubochet, J., Baschong, W., Kellenberger, E.: Organization of
         double-stranded DNA in bacteriophages: a study by cryo-electron microscopy
         of vitrified samples. EMBO J. **6**, 1507–1512 (1987).

[HMC]   Hass, R., Murphy, R.F., Cantor, C.R.: Testing models of the arrangement of
         DNA inside bacteriophage λ by crosslinking the packaged DNA. J. Mol. Biol.
         159, 71–92 (1982).

[Ser]   Serwer, P.: Arrangement of double-stranded DNA packaged in bacteriophage
         capsids : An alternative model. J. Mol. Biol. **190**, 509–512 (1986).

[RWC]   Richards, K., Williams, R., Calendar, R.: Mode of DNA packing within bacte-
         riophage heads. J. Mol. Biol. **78**, 255–259 (1973).

[CCR]   Cerritelli, M., Cheng, N., Rosenberg, A., McPherson, C., Booy, F., Steven, A.:
         Encapsidated Conformation of Bacteriophage T7 DNA. Cell **91**, 271–280 (1997).

[BNB]   Black, L., Newcomb, W., Boring, J., Brown, J.: Ion Etching of Bacteriophage T4:
         Support for a Spiral-Fold Model of Packaged DNA. Proc. Natl. Acad. Sci. USA
         82, 7960–7964 (1985).

[Hud]   Hud, N.: Double-stranded DNA organization in bacteriophage heads: an alterna-
         tive toroid-based model. Biophys. J. **69**, 1355–1362 (1995).

[JCJ]   Jiang, W., Chang, J., Jakana, J., Weigele, P., King, J., Chiu, W.: Strucure
         of epsilon15 bacteriophage reveals genome organization and DNA packag-
         ing/injection apparatus. Nature **439**, 612–616 (2006).

[WMC]   Wang, J.C., Martin, K.V., Calendar, R.: Sequence similarity of the cohesive ends
         of coliphage P4, P2, and 186 deoxyribonucleic acid. Biochemistry **12**, 2119–2123
         (1973).

[LDC]   Liu, L.F., Davis, J.L., Calendar, R.: Novel topologically knotted DNA from bac-
         teriophage P4 capsids: studies with DNA topoisomerases. Nucleic Acids Res. **9**,
         3979–3989 (1981).

[WMH]   Wolfson, J.S., McHugh, G.L., Hooper, D.C., Swartz, M.N.: Knotting of DNA
         molecules isolated from deletion mutants of intact bacteriophage P4. Nucleic
         Acids Res. **13**, 6695–6702 (1985).

[IJC] Isaken, M., Julien, B., Calendar, R., Lindgvist, B.H.: Isolation of knotted DNA from coliphage P4: in DNA Topoisomerase Protocols, DNA Topology, and Enzymes, eds. Bjornsti, M. A. & Osheroff, N. (Humana, Totowa, NJ), Vol. **94**, pp. 69–74 (1999).

[RDM] Raimondi, A. Donghi, R., Montaguti, A., Pessina, A., Deho, G.: Analysis of spontaneous deletion mutants of satellite bacteriophage P4. J. Virol. **54**, 233–235 (1985).

[RUV] Rybenkov, V.V., Ullsperger, C., Vologodskii, A.V., Cozarelli, N.R.: Simplification of DNA Topology Below Equilibrium Values by Type II Topoisomerases. Science **277**, 690–693 (1997).

[MRR] Metropolis, N., Rosenbluth, A.W., Rosenbluth, M.N., Teller, A.H., Teller, E.: Equation of State Calculations by Fast Computing Machines. J. Chem. Phys. **21**, 1087–1092 (1953).

[Mil] Millet, K.: Knotting of Regular Polygons in 3-Space. In Series of Knots and Everything, eds. Sumners D.W. & Millet K.C. (World Scientific, Singapore), Vol. **7**, pp. 31–46 (1994).

[MS]   Madras, N., Slade, G.: The Self-Avoiding Walk, Birkhauser, Boston (1993).

[DT2] Deguchi, T., Tsurusaki, K.: A Statistical Study of Random Knotting Using the Vassiliev Invariants. J. Knot Theor. Ramifications **3**, 321–353 (1994).

[SKB] Stasiak, A., Katrich, V., Bednar, J., Michoud, D., Dubochet, J.: Electrophoretic mobility of DNA knots. Nature **384**, 122 (1996).

[VCL] Vologodskii, A.V., Crisona, N.J., Laurie, B., Pieranski, P., Katritch, V., Dubochet, J., Stasiak, A.: Sedimentation and electrophoretic migration of DNA knots and catenanes. J. Mol. Biol. **278**, 1–3 (1998).

[KKS] Kanaar, R., Klippel, A., Shekhtman, E., Dungan, J., Kahmann, R., Cozzarelli, N.R.: Processive recombination by the phage Mu Gin system: Implications for the mechanisms of DNA strand exchange, DNA site alignment, and enhancer action. Cell **62**, 353–366 (1990).

[WSR] Weber, C., Stasiak, A., De Los Rios, P., Dietler, G.: Numerical Simulation of Gel Electrophoresis of DNA Knots in Weak and Strong Electric Fields. Biophys. J. **90**, 3100–3105 (2006).

[RSW] van Rensburg, J., Sumners, D.W., Whittington, S.G.: The Writhe of Knots and Links, In Ideal Knots, Series on Knots and Everything, eds. Stasiak, A., Katritch, V., Kauffman, L.H. (World Scientific, Singapore), Vol. **19**, 70–87 (1999).

[MM]   Marenduzzo, D., Micheletti, C.: Thermodynamics of DNA packaging inside a viral capsid: The role of DNA intrinsic thickness. J. Mol. Biol. **330**, 485–492 (2003).

[MMT] Maritan, A., Micheletti, C., Trovato, A., Banavar, J.: Optimal shapes of compact strings. Nature **406**, 287–289 (2000).

[TK]   Tzil, S., Kindt, J.T., Gelbart, W., Ben-Shaul, A.: Nucleic acid packaging of DNA viruses. Biophys. J. **84**, 1616–1627 (2003).

[LL]   LaMarque, J.C., Le, T.L., Harvey, S.C.: Packaging double-helical DNA into viral capsids. Biopolymers **73**, 3480–355 (2004).

Index

LIST OF C.I.M.E. SEMINARS

Published by C.I.M.E

Published by Ed. Cremonese, Firenze

Published by Ed. Liguori, Napoli

Published by Ed. Liguori, Napoli & Birkhäuser

Published by Springer-Verlag

Lecture Notes in Mathematics

For information about earlier volumes
please contact your bookseller or Springer
LNM Online archive: springerlink.com

Vol. 1830: M. I. Gil', Operator Functions and Localization of Spectra. XIV, 256 p, 2003.

Vol. 1831: A. Connes, J. Cuntz, E. Guentner, N. Higson, J. E. Kaminker, Noncommutative Geometry, Martina Franca, Italy 2002. Editors: S. Doplicher, L. Longo (2004)

Vol. 1832: J. Azéma, M. Émery, M. Ledoux, M. Yor (Eds.), Séminaire de Probabilités XXXVII (2003)

Vol. 1833: D.-Q. Jiang, M. Qian, M.-P. Qian, Mathematical Theory of Nonequilibrium Steady States. On the Frontier of Probability and Dynamical Systems. IX, 280 p, 2004.

Vol. 1834: Yo. Yomdin, G. Comte, Tame Geometry with Application in Smooth Analysis. VIII, 186 p, 2004.

Vol. 1835: O.T. Izhboldin, B. Kahn, N.A. Karpenko, A. Vishik, Geometric Methods in the Algebraic Theory of Quadratic Forms. Summer School, Lens, 2000. Editor: J.-P. Tignol (2004)

Vol. 1836: C. Năstăsescu, F. Van Oystaeyen, Methods of Graded Rings. XIII, 304 p, 2004.

Vol. 1837: S. Tavaré, O. Zeitouni, Lectures on Probability Theory and Statistics. Ecole d'Eté de Probabilités de Saint-Flour XXXI-2001. Editor: J. Picard (2004)

Vol. 1838: A.J. Ganesh, N.W. O'Connell, D.J. Wischik, Big Queues. XII, 254 p, 2004.

Vol. 1839: R. Gohm, Noncommutative Stationary Processes. VIII, 170 p, 2004.

Vol. 1840: B. Tsirelson, W. Werner, Lectures on Probability Theory and Statistics. Ecole d'Eté de Probabilités de Saint-Flour XXXII-2002. Editor: J. Picard (2004)

Vol. 1841: W. Reichel, Uniqueness Theorems for Variational Problems by the Method of Transformation Groups (2004)

Vol. 1842: T. Johnsen, A. L. Knutsen, K_3 Projective Models in Scrolls (2004)

Vol. 1843: B. Jefferies, Spectral Properties of Noncommuting Operators (2004)

Vol. 1844: K.F. Siburg, The Principle of Least Action in Geometry and Dynamics (2004)

Vol. 1845: Min Ho Lee, Mixed Automorphic Forms, Torus Bundles, and Jacobi Forms (2004)

Vol. 1846: H. Ammari, H. Kang, Reconstruction of Small Inhomogeneities from Boundary Measurements (2004)

Vol. 1847: T.R. Bielecki, T. Björk, M. Jeanblanc, M. Rutkowski, J.A. Scheinkman, W. Xiong, Paris-Princeton Lectures on Mathematical Finance 2003 (2004)

Vol. 1848: M. Abate, J. E. Fornaess, X. Huang, J. P. Rosay, A. Tumanov, Real Methods in Complex and CR Geometry, Martina Franca, Italy 2002. Editors: D. Zaitsev, G. Zampieri (2004)

Vol. 1849: Martin L. Brown, Heegner Modules and Elliptic Curves (2004)

Vol. 1850: V. D. Milman, G. Schechtman (Eds.), Geometric Aspects of Functional Analysis. Israel Seminar 2002-2003 (2004)

Vol. 1851: O. Catoni, Statistical Learning Theory and Stochastic Optimization (2004)

Vol. 1852: A.S. Kechris, B.D. Miller, Topics in Orbit Equivalence (2004)

Vol. 1853: Ch. Favre, M. Jonsson, The Valuative Tree (2004)

Vol. 1854: O. Saeki, Topology of Singular Fibers of Differential Maps (2004)

Vol. 1855: G. Da Prato, P.C. Kunstmann, I. Lasiecka, A. Lunardi, R. Schnaubelt, L. Weis, Functional Analytic Methods for Evolution Equations. Editors: M. Iannelli, R. Nagel, S. Piazzera (2004)

Vol. 1856: K. Back, T.R. Bielecki, C. Hipp, S. Peng, W. Schachermayer, Stochastic Methods in Finance, Bressanone/Brixen, Italy, 2003. Editors: M. Fritelli, W. Runggaldier (2004)

Vol. 1857: M. Émery, M. Ledoux, M. Yor (Eds.), Séminaire de Probabilités XXXVIII (2005)

Vol. 1858: A.S. Cherny, H.-J. Engelbert, Singular Stochastic Differential Equations (2005)

Vol. 1859: E. Letellier, Fourier Transforms of Invariant Functions on Finite Reductive Lie Algebras (2005)

Vol. 1860: A. Borisyuk, G.B. Ermentrout, A. Friedman, D. Terman, Tutorials in Mathematical Biosciences I. Mathematical Neurosciences (2005)

Vol. 1861: G. Benettin, J. Henrard, S. Kuksin, Hamiltonian Dynamics – Theory and Applications, Cetraro, Italy, 1999. Editor: A. Giorgilli (2005)

Vol. 1862: B. Helffer, F. Nier, Hypoelliptic Estimates and Spectral Theory for Fokker-Planck Operators and Witten Laplacians (2005)

Vol. 1863: H. Führ, Abstract Harmonic Analysis of Continuous Wavelet Transforms (2005)

Vol. 1864: K. Efstathiou, Metamorphoses of Hamiltonian Systems with Symmetries (2005)

Vol. 1865: D. Applebaum, B.V. R. Bhat, J. Kustermans, J. M. Lindsay, Quantum Independent Increment Processes I. From Classical Probability to Quantum Stochastic Calculus. Editors: M. Schürmann, U. Franz (2005)

Vol. 1866: O.E. Barndorff-Nielsen, U. Franz, R. Gohm, B. Kümmerer, S. Thorbjønsen, Quantum Independent Increment Processes II. Structure of Quantum Lévy Processes, Classical Probability, and Physics. Editors: M. Schürmann, U. Franz, (2005)

Vol. 1867: J. Sneyd (Ed.), Tutorials in Mathematical Biosciences II. Mathematical Modeling of Calcium Dynamics and Signal Transduction. (2005)

Vol. 1868: J. Jorgenson, S. Lang, $Pos_n(R)$ and Eisenstein Series. (2005)

Vol. 1869: A. Dembo, T. Funaki, Lectures on Probability Theory and Statistics. Ecole d'Eté de Probabilités de Saint-Flour XXXIII-2003. Editor: J. Picard (2005)

Vol. 1870: V.I. Gurariy, W. Lusky, Geometry of Müntz Spaces and Related Questions. (2005)

Vol. 1871: P. Constantin, G. Gallavotti, A.V. Kazhikhov, Y. Meyer, S. Ukai, Mathematical Foundation of Turbulent Viscous Flows, Martina Franca, Italy, 2003. Editors: M. Cannone, T. Miyakawa (2006)

Vol. 1872: A. Friedman (Ed.), Tutorials in Mathematical Biosciences III. Cell Cycle, Proliferation, and Cancer (2006)

Vol. 1873: R. Mansuy, M. Yor, Random Times and Enlargements of Filtrations in a Brownian Setting (2006)

Vol. 1874: M. Yor, M. Émery (Eds.), In Memoriam Paul-André Meyer - Séminaire de Probabilités XXXIX (2006)

Vol. 1875: J. Pitman, Combinatorial Stochastic Processes. Ecole d'Eté de Probabilités de Saint-Flour XXXII-2002. Editor: J. Picard (2006)

Vol. 1876: H. Herrlich, Axiom of Choice (2006)

Vol. 1877: J. Steuding, Value Distributions of L-Functions (2007)

Vol. 1878: R. Cerf, The Wulff Crystal in Ising and Percolation Models, Ecole d'Eté de Probabilités de Saint-Flour XXXIV-2004. Editor: Jean Picard (2006)

Vol. 1879: G. Slade, The Lace Expansion and its Applications, Ecole d'Eté de Probabilités de Saint-Flour XXXIV-2004. Editor: Jean Picard (2006)

Vol. 1880: S. Attal, A. Joye, C.-A. Pillet, Open Quantum Systems I, The Hamiltonian Approach (2006)

Vol. 1881: S. Attal, A. Joye, C.-A. Pillet, Open Quantum Systems II, The Markovian Approach (2006)

Vol. 1882: S. Attal, A. Joye, C.-A. Pillet, Open Quantum Systems III, Recent Developments (2006)

Vol. 1883: W. Van Assche, F. Marcellàn (Eds.), Orthogonal Polynomials and Special Functions, Computation and Application (2006)

Vol. 1884: N. Hayashi, E.I. Kaikina, P.I. Naumkin, I.A. Shishmarev, Asymptotics for Dissipative Nonlinear Equations (2006)

Vol. 1885: A. Telcs, The Art of Random Walks (2006)

Vol. 1886: S. Takamura, Splitting Deformations of Degenerations of Complex Curves (2006)

Vol. 1887: K. Habermann, L. Habermann, Introduction to Symplectic Dirac Operators (2006)

Vol. 1888: J. van der Hoeven, Transseries and Real Differential Algebra (2006)

Vol. 1889: G. Osipenko, Dynamical Systems, Graphs, and Algorithms (2006)

Vol. 1890: M. Bunge, J. Funk, Singular Coverings of Toposes (2006)

Vol. 1891: J.B. Friedlander, D.R. Heath-Brown, H. Iwaniec, J. Kaczorowski, Analytic Number Theory, Cetraro, Italy, 2002. Editors: A. Perelli, C. Viola (2006)

Vol. 1892: A. Baddeley, I. Bárány, R. Schneider, W. Weil, Stochastic Geometry, Martina Franca, Italy, 2004. Editor: W. Weil (2007)

Vol. 1893: H. Hanßmann, Local and Semi-Local Bifurcations in Hamiltonian Dynamical Systems, Results and Examples (2007)

Vol. 1894: C.W. Groetsch, Stable Approximate Evaluation of Unbounded Operators (2007)

Vol. 1895: L. Molnár, Selected Preserver Problems on Algebraic Structures of Linear Operators and on Function Spaces (2007)

Vol. 1896: P. Massart, Concentration Inequalities and Model Selection, Ecole d'Été de Probabilités de Saint-Flour XXXIII-2003. Editor: J. Picard (2007)

Vol. 1897: R. Doney, Fluctuation Theory for Lévy Processes, Ecole d'Été de Probabilités de Saint-Flour XXXV-2005. Editor: J. Picard (2007)

Vol. 1898: H.R. Beyer, Beyond Partial Differential Equations, On linear and Quasi-Linear Abstract Hyperbolic Evolution Equations (2007)

Vol. 1899: Séminaire de Probabilités XL. Editors: C. Donati-Martin, M. Émery, A. Rouault, C. Stricker (2007)

Vol. 1900: E. Bolthausen, A. Bovier (Eds.), Spin Glasses (2007)

Vol. 1901: O. Wittenberg, Intersections de deux quadriques et pinceaux de courbes de genre 1, Intersections of Two Quadrics and Pencils of Curves of Genus 1 (2007)

Vol. 1902: A. Isaev, Lectures on the Automorphism Groups of Kobayashi-Hyperbolic Manifolds (2007)

Vol. 1903: G. Kresin, V. Maz'ya, Sharp Real-Part Theorems (2007)

Vol. 1904: P. Giesl, Construction of Global Lyapunov Functions Using Radial Basis Functions (2007)

Vol. 1905: C. Prévôt, M. Röckner, A Concise Course on Stochastic Partial Differential Equations (2007)

Vol. 1906: T. Schuster, The Method of Approximate Inverse: Theory and Applications (2007)

Vol. 1907: M. Rasmussen, Attractivity and Bifurcation for Nonautonomous Dynamical Systems (2007)

Vol. 1908: T.J. Lyons, M. Caruana, T. Lévy, Differential Equations Driven by Rough Paths, Ecole d'Été de Probabilités de Saint-Flour XXXIV-2004 (2007)

Vol. 1909: H. Akiyoshi, M. Sakuma, M. Wada, Y. Yamashita, Punctured Torus Groups and 2-Bridge Knot Groups (I) (2007)

Vol. 1910: V.D. Milman, G. Schechtman (Eds.), Geometric Aspects of Functional Analysis. Israel Seminar 2004-2005 (2007)

Vol. 1911: A. Bressan, D. Serre, M. Williams, K. Zumbrun, Hyperbolic Systems of Balance Laws. Cetraro, Italy 2003. Editor: P. Marcati (2007)

Vol. 1912: V. Berinde, Iterative Approximation of Fixed Points (2007)

Vol. 1913: J.E. Marsden, G. Misiołek, J.-P. Ortega, M. Perlmutter, T.S. Ratiu, Hamiltonian Reduction by Stages (2007)

Vol. 1914: G. Kutyniok, Affine Density in Wavelet Analysis (2007)

Vol. 1915: T. Bıyıkoğlu, J. Leydold, P.F. Stadler, Laplacian Eigenvectors of Graphs. Perron-Frobenius and Faber-Krahn Type Theorems (2007)

Vol. 1916: C. Villani, F. Rezakhanlou, Entropy Methods for the Boltzmann Equation. Editors: F. Golse, S. Olla (2008)

Vol. 1917: I. Veselić, Existence and Regularity Properties of the Integrated Density of States of Random Schrödinger (2008)

Vol. 1918: B. Roberts, R. Schmidt, Local Newforms for GSp(4) (2007)

Vol. 1919: R.A. Carmona, I. Ekeland, A. Kohatsu-Higa, J.-M. Lasry, P.-L. Lions, H. Pham, E. Taflin, Paris-Princeton Lectures on Mathematical Finance 2004. Editors: R.A. Carmona, E. Çinlar, I. Ekeland, E. Jouini, J.A. Scheinkman, N. Touzi (2007)

Vol. 1920: S.N. Evans, Probability and Real Trees. Ecole d'Été de Probabilités de Saint-Flour XXXV-2005 (2008)

Vol. 1921: J.P. Tian, Evolution Algebras and their Applications (2008)

Vol. 1922: A. Friedman (Ed.), Tutorials in Mathematical BioSciences IV. Evolution and Ecology (2008)

Vol. 1923: J.P.N. Bishwal, Parameter Estimation in Stochastic Differential Equations (2008)

Vol. 1924: M. Wilson, Littlewood-Paley Theory and Exponential-Square Integrability (2008)

Vol. 1925: M. du Sautoy, L. Woodward, Zeta Functions of Groups and Rings (2008)

Vol. 1926: L. Barreira, V. Claudia, Stability of Nonautonomous Differential Equations (2008)

Vol. 1927: L. Ambrosio, L. Caffarelli, M.G. Crandall, L.C. Evans, N. Fusco, Calculus of Variations and Non-Linear Partial Differential Equations. Cetraro, Italy 2005. Editors: B. Dacorogna, P. Marcellini (2008)

Vol. 1928: J. Jonsson, Simplicial Complexes of Graphs (2008)

Vol. 1929: Y. Mishura, Stochastic Calculus for Fractional Brownian Motion and Related Processes (2008)

Vol. 1930: J.M. Urbano, The Method of Intrinsic Scaling. A Systematic Approach to Regularity for Degenerate and Singular PDEs (2008)

Vol. 1931: M. Cowling, E. Frenkel, M. Kashiwara, A. Valette, D.A. Vogan, Jr., N.R. Wallach, Representation Theory and Complex Analysis. Venice, Italy 2004. Editors: E.C. Tarabusi, A. D'Agnolo, M. Picardello (2008)

Vol. 1932: A.A. Agrachev, A.S. Morse, E.D. Sontag, H.J. Sussmann, V.I. Utkin, Nonlinear and Optimal

Recent Reprints and New Editions

LECTURE NOTES IN MATHEMATICS   🦄 Springer

Edited by J.-M. Morel, F. Takens, B. Teissier, P.K. Maini

Editorial Policy (for the publication of monographs)

1. Lecture Notes aim to report new developments in all areas of mathematics and their applications - quickly, informally and at a high level. Mathematical texts analysing new developments in modelling and numerical simulation are welcome.

   Monograph manuscripts should be reasonably self-contained and rounded off. Thus they may, and often will, present not only results of the author but also related work by other people. They may be based on specialised lecture courses. Furthermore, the manuscripts should provide sufficient motivation, examples and applications. This clearly distinguishes Lecture Notes from journal articles or technical reports which normally are very concise. Articles intended for a journal but too long to be accepted by most journals, usually do not have this "lecture notes" character. For similar reasons it is unusual for doctoral theses to be accepted for the Lecture Notes series, though habilitation theses may be appropriate.

2. Manuscripts should be submitted either online at www.editorialmanager.com/lnm to Springer's mathematics editorial in Heidelberg, or to one of the series editors. In general, manuscripts will be sent out to 2 external referees for evaluation. If a decision cannot yet be reached on the basis of the first 2 reports, further referees may be contacted: The author will be informed of this. A final decision to publish can be made only on the basis of the complete manuscript, however a refereeing process leading to a preliminary decision can be based on a pre-final or incomplete manuscript. The strict minimum amount of material that will be considered should include a detailed outline describing the planned contents of each chapter, a bibliography and several sample chapters.

   Authors should be aware that incomplete or insufficiently close to final manuscripts almost always result in longer refereeing times and nevertheless unclear referees' recommendations, making further refereeing of a final draft necessary.

   Authors should also be aware that parallel submission of their manuscript to another publisher while under consideration for LNM will in general lead to immediate rejection.

3. Manuscripts should in general be submitted in English. Final manuscripts should contain at least 100 pages of mathematical text and should always include

   – a table of contents;
   – an informative introduction, with adequate motivation and perhaps some historical remarks: it should be accessible to a reader not intimately familiar with the topic treated;
   – a subject index: as a rule this is genuinely helpful for the reader.

   For evaluation purposes, manuscripts may be submitted in print or electronic form (print form is still preferred by most referees), in the latter case preferably as pdf- or zipped ps-files. Lecture Notes volumes are, as a rule, printed digitally from the authors' files. To ensure best results, authors are asked to use the LaTeX2e style files available from Springer's web-server at:

   ftp://ftp.springer.de/pub/tex/latex/svmonot1/ (for monographs) and
   ftp://ftp.springer.de/pub/tex/latex/svmultt1/ (for summer schools/tutorials).

Additional technical instructions, if necessary, are available on request from: lnm@springer.com.

4. Careful preparation of the manuscripts will help keep production time short besides ensuring satisfactory appearance of the finished book in print and online. After acceptance of the manuscript authors will be asked to prepare the final LaTeX source files and also the corresponding dvi-, pdf- or zipped ps-file. The LaTeX source files are essential for producing the full-text online version of the book (see http://www.springerlink.com/openurl.asp?genre=journal&issn=0075-8434 for the existing online volumes of LNM).

The actual production of a Lecture Notes volume takes approximately 12 weeks.

5. Authors receive a total of 50 free copies of their volume, but no royalties. They are entitled to a discount of 33.3% on the price of Springer books purchased for their personal use, if ordering directly from Springer.

6. Commitment to publish is made by letter of intent rather than by signing a formal contract. Springer-Verlag secures the copyright for each volume. Authors are free to reuse material contained in their LNM volumes in later publications: a brief written (or e-mail) request for formal permission is sufficient.

Addresses:
Professor J.-M. Morel, CMLA,
École Normale Supérieure de Cachan,
61 Avenue du Président Wilson, 94235 Cachan Cedex, France
E-mail: Jean-Michel.Morel@cmla.ens-cachan.fr

Professor F. Takens, Mathematisch Instituut,
Rijksuniversiteit Groningen, Postbus 800,
9700 AV Groningen, The Netherlands
E-mail: F.Takens@rug.nl

Professor B. Teissier, Institut Mathématique de Jussieu,
UMR 7586 du CNRS, Équipe "Géométrie et Dynamique",
175 rue du Chevaleret,
75013 Paris, France
E-mail: teissier@math.jussieu.fr

For the "Mathematical Biosciences Subseries" of LNM:

Professor P.K. Maini, Center for Mathematical Biology,
Mathematical Institute, 24-29 St Giles,
Oxford OX1 3LP, UK
E-mail: maini@maths.ox.ac.uk

Springer, Mathematics Editorial, Tiergartenstr. 17,
69121 Heidelberg, Germany,
Tel.: +49 (6221) 487-259
Fax: +49 (6221) 4876-8259
E-mail: lnm@springer.com